# Cell Biology Monographs
## Continuation of Protoplasmatologia

Vol. 2

Springer-Verlag
New York Wien

# The Golgi Apparatus

W. G. Whaley

Springer-Verlag
New York Wien

W. GORDON WHALEY

Ashbel Smith Professor of Cellular Biology
The Cell Research Institute
The University of Texas at Austin
Austin, Texas, U.S.A.

*574.8734*
*W55g*
*100216*
*Mar. 1977*

© 1975 by Springer-Verlag/Wien
Printed in Austria by Adolf Holzhausens Nfg., Wien

## With 97 Figures

Library of Congress Cataloging in Publication Data. Whaley, William Gordon, 1914—. The Golgi apparatus.
(Cell biology monographs; v. 2.) Bibliography: p. 1. Golgi apparatus. I. Title. II. Series. [DNLM: Golgi apparatus.
2. Physiology. W1. CE128H v. 2 / QH603.G6 W137g] QH603.G6W48 574.8'734, 75-20055.

ISBN 0-387-81315-2 Springer-Verlag New York-Wien
ISBN 3-211-81315-2 Springer-Verlag Wien-New York

*Dedicated to*
*Wesley West*
*a gentleman with pioneer spirit*

About twenty years ago, Prof. FRIEDL WEBER (Graz University) and Prof. L. V. HEILBRUNN (University of Pennsylvania) conceived the idea for the handbook "Protoplasmatologia" at a time when the state of knowledge in the field of cell biology still permitted one to think of an all-encompassing handbook in the classical sense. Since 1953 fifty-three volumes with a total of about 9,500 pages have been published. The very rapid developments in this area of science, especially during the last decade, have led to new insights which necessitated some alterations in the original plan of the handbook; also, changes in the board of editors since the death of the founders have brought about a reorientation of viewpoints.

The editors, in agreement with the publisher, have decided to abandon the confining limits of the original disposition of the handbook altogether and to continue this work, in a form more appropriate to current needs, as an open series of monographs dealing with present-day problems and findings in cell biology. This will make it possible to treat the most modern and interesting aspects of the field as they arise in the course of contemporary research. The highest scientific, editorial and publishing standards will continue to be maintained.

Editors and publisher

# Preface

A comprehensive review of the Golgi apparatus and its functioning would require a multi-volume publication and not a monograph and it would be so repetitious as to discourage the reader. The requirement at this stage is for a reinterpretation of the character and functioning of this organelle since the last major interpretations have concentrated on its role in secretion and it has now been shown to be a component of essentially all cells whether or not they have been traditionally emphasized as secreting cells. As a consequence the efforts have been placed on the common characteristics of the organelle, a postulate concerning its functioning in cells generally, and the details of variations where these seem important.

The major acknowledgment of assistance in compiling the material must go to the investigators whose contributions, sometimes positive and sometimes of a character to spur additional investigations, allowed the development of this postulate. The paper has been prepared with the detailed assistance of Dr. MARIANNE DAUWALDER who, by her own studies and her insight into the significance of other studies, has been a working partner of many years in the development of a general hypothesis and whose knowledge of investigations of the Golgi apparatus is great enough to let her call attention to instances of support and contention with the general functional hypothesis that has been involved.

Thanks are due to Dr. AUDREY NELSON SLATE for concern with the details of the dusty, older literature which occasionally contributes to the progressive development of ideas that have been incorporated into modern interpretations. Technical assistance in the selection and preparation of the micrographs for publication has been rendered by Miss JOYCE E. KEPHART and in photographic reproduction by Mr. GERALD FAY. Assistance with manuscript preparation has been given by KATHIE DODDS, ROY W. HOLLEY, and CLARE Y. WHALEY.

The collection of original data presented in the paper has been supported by NSF Research Grant #17778. Both new findings and time for their interpretation have been made possible by support from the Faith Foundation.

Finally, thanks must be expressed for the extended patience of Dr. MAX ALFERT of the Department of Zoology, University of California at Berkeley, and Dr. W. SCHWABL of the publisher's office in Vienna.

The work has been a long time in preparation but the author offers no apologies for this since had it been produced earlier it could have been only one more addition to the reviews dependent on earlier interpretations whereas now it can be an interpretation of a general functioning of the organelle within the economy of the cell. If the interpretation is correct or even partly correct it allows for an understanding of a continuity of genetic control extending from the genome to extracellular activities that are important in the development, association, function, dysfunctions, and perhaps even other aspects in the behavior of cells.

Austin, Texas, January 1975                              W. GORDON WHALEY

# Contents

# I. Introduction

It is three-quarters of a century since what is now called the Golgi apparatus was first reported yet there are still unanswered questions concerning it. CAMILLO GOLGI put his observations into the literature under the term *apparato reticolare interno* (internal reticular apparatus) in 1898 (Fig. 1).

An extended historical review of the Golgi apparatus is unnecessary because most of the original papers are still available, and older interpretations made

Fig. 1. One of GOLGI's original drawings of the internal reticular apparatus as seen in a Purkinje cell of a barn owl. From Arch. ital. Biol. **30**, 1898.

with less than satisfactory techniques and with the limitations of light microscopy have been largely superseded. Further, there has been a series of excellent reviews and interpretations, which, if read in sequence, gives a good picture of the successive developments (DUESBERG 1914, CAJAL 1914, COWDRY 1923, 1924, BOWEN 1926, 1929, KIRKMAN and SEVERINGHAUS 1938 a, b, and c, BEAMS and KESSEL 1968, MORRÉ *et al.* 1971, DAUWALDER *et al.* 1972, COOK 1973. See also a symposium held by the Royal Microscopical Society in 1954 [1955]). There are, however, several reasons for including a brief historical background in a work devoted to the Golgi apparatus. Studies of the organelle by continuously improved techniques and combinations of techniques provide a good illustration of the advancement of biological science from the observational stage to interpretations of function. They may possibly lead to modification of the organelle and its function as a means of improving the state of the organism. The ubiquitousness of the apparatus calls for consideration of a general cellular activity on the part of a component that was once looked upon by many investigators as a curiosity characteristic of only certain cells and then considered only in terms of specific activities.

No cellular organelle has been the subject of as many, as long-lasting, or as diverse polemics as the Golgi apparatus. They have been repeatedly reviewed and little is to be gained by a comprehensive reconsideration of the grounds for them or the confusion to which they led. It is now clear that all but a few eukaryotic cells have Golgi apparatus with common elements of structure and function. The organelle is extremely pleiomorphic, subject to changes in position, intensity of activity, and the character of its products. Moreover, it may perform several functions simultaneously. There is, however, a core of recognizable characteristics. Because there is, one must treat even the modern literature selectively lest the treatment become repetitive or a mere cataloguing of activities that occur principally in the same pattern. For this reason, the concentration here will be on the basic aspects of the pattern, instances where it is notably related to other cellular functions, and finally the interpretations that can be made from its activities.

## II. Early Studies

### A. Pre-Golgi Studies

One of the functions of the Golgi apparatus is participation in the formation of the acrosome in the development of the sperm. This activity was fairly well worked out before the general occurrence of the organelle was recognized.

The structure now known as the Golgi apparatus was apparently discovered by LA VALETTE ST. GEORGE (1865, 1867). After GOLGI's reporting of the organelle in 1898, PERRONCITO (1910) connected it with LA VALETTE ST. GEORGE's observations. DOUGLAS (1935) wrote a paper on LA VALETTE ST. GEORGE's discovery of the Golgi apparatus and the mitochondria. GATENBY (1955) also gave credit to him for the discovery of the Golgi apparatus as did BEAMS and KESSEL (1968). A study of LA VALETTE ST. GEORGE's paper makes it difficult to support the idea that he saw the organelle clearly.

WILSON (1925) reported that the details of *dictyokinesis*, a term used by PERRONCITO in 1910 to describe the division of the Golgi apparatus, had been clearly described by PLATNER in 1889 in the spermatocytes of *Helix* and by MURRAY (1898) in those of *Helix* and *Arion*. Both these observations were made before GOLGI's paper of 1898 was known. PLATNER derived the dictyosomes from a *Nebenkern* (Fig. 2) and MURRAY from an *attraction sphere* (Fig. 3). Both of the structures observed were subsequently shown to be homologous to the Golgi apparatus. Some of these earlier pictures depict the distribution of elements in dividing cells vividly. Much later (1923) CAJAL claimed he had discovered the *endocellular apparatus* (a term sometimes applied to the Golgi apparatus) in muscle cells of insects in 1890. He said he did not record it because he was not convinced that he had found a new organelle.

Considering the fixatives and stains being used, the amount of experimental work with them characteristic of the latter part of the 19th century, and the multiplicity of the tissue and cell types being studied, it seems reasonable to suggest that a considerable number of investigators may have seen this

Fig. 2. A drawing of PLATNER showing a Nebenkern adjacent to the nucleus. From Arch. mikrosk. Anat. **33**, 1889.

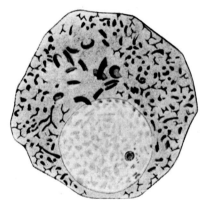

Fig. 3. The division of an attraction sphere in *Helix pomatia* according to MURRAY. By the original interpretation the attraction sphere breaks up into rods, the number of which is constant and equal to half the chromosome number. In this diagram paired rods are distinguished by shading and appear just above the nucleus. From Zool. Jb. **11**, 1898.

pleiomorphic organelle before GOLGI. It was GOLGI, however, who devised a method that put the organelle in sharp contrast with other cellular components and permitted him and his students to demonstrate it as a consistent structural component in a wide variety of tissue cell types. It was also GOLGI who recognized not only some of the details of its structure but also that it is variable in character and position.

## B. Camillo Golgi

CAMILLO GOLGI was born on July 9, 1843 in Corteno, Brescia, Lombardy. He was the son of a physician who took up a practice in a village near Pavia. It is generally assumed that this fact had some impact on GOLGI's decision to study medicine at the University of Pavia. He entered the University, which had a strong tradition and a distinguished faculty in biological science and medicine. In a summary of GOLGI's work at the time of GOLGI's death, DA FANO (1926) pointed out that biological and medical studies were undergoing a considerable renewal in Italy at the time; they made a particularly deep impression on the young student. GOLGI received his medical degree in 1865 and began a career as clinician, researcher, and teacher that lasted until 1926. It was a career monumental in its achievements.

After graduation he accepted a position in the Ospedale di San Matteo in Pavia. He served in this position from 1865 to 1871 while at the same time continuing to work in various sections of the University. Of particular importance to his career was a new laboratory of experimental pathology which had been established by PAOLO MANTEGAZZA and had come under the direction of GIULIO BIZZOZERO with whom GOLGI had close and profitable contact. Later, under the name of the Institute of General Pathology and Histology, this laboratory was headed by GOLGI himself and it is now named after him.

He also came under another influence in the person of CESARE LOMBROSO, the author of *L'uomo deliquente,* who had a deep interest in possible relationships between individual development and criminal tendencies. DA FANO (1926) has pointed out that the two men were so different temperamentally that the liaison did not last long, but some of GOLGI's early work seems to bear the stamp of LOMBROSO's influence as did his continuing interest in the structure and working of the central nervous system. His interests in pathology and histology became increasingly deeply established.

In 1872, GOLGI left Pavia to become resident physician of the Home for Incurables in nearby Abbiategrasso, but he retained his interest in research. Here in 1873 he discovered his chromate of silver method (*la reazione nera*) with which in later years he was to revolutionize the study of the nervous system. Other investigators were to pick up this technique and with various refinements take some of these studies far beyond where GOLGI himself took them. GOLGI returned to Pavia in 1875 and except for a short period at the University of Siena remained there and continued to be highly productive in research and teaching until nearly the end of his life in 1926. GOLGI began to publish in the 1860's and continued almost until the time of his death, frequently turning out as many as eight papers a year. His early works are concerned with cellular and tissue developments and in some instances their modification in pathological conditions.

After his return to Pavia, GOLGI continued a series of publications on the nervous system occupying himself from 1882 to 1885 with a publication of an eight-part work *Sulla fina anatomia degli organi centrali del sistemi nervosa* which was reprinted in French and German and was one of the great compilations of the time. Besides many histological works during this period and subsequently, some of which were related to changes in the nervous system, he published a number of experimental papers on various disease conditions—perhaps the most important of which was his work on malaria. Done in great detail, it established definitely the character of the malaria organism, the relation between the life cycle of the organism and fever, and the effectiveness of quinine as a treatment. Had he not made significant enough contributions in other fields, his malaria work alone would have withstood the test of time. DA FANO called attention to the fact that in 1926 some of GOLGI's papers were still workable models, and 50 years later, some are still important.

GOLGI's papers were republished in four volumes: *Opera Omnia* (1903 to 1929). Volumes 1 and 2 contain his histological papers from 1870—1902; volume 3 is his pathological papers 1868—1894; and volume 4 contains papers published between 1903 and 1925.

## C. La Reazione Nera

Of particular importance for present considerations is *la reazione nera,* or as it came to be known later, Golgi's method. The reagents used by histologists in GOLGI's time were primarily potassium bichromate, osmic acid, and silver nitrate and it was with these that GOLGI was experimenting. The details of GOLGI's discovery are not clear. Fundamentally it consisted of placing tissues in a silver nitrate bath after a preliminary fixation in bichromate solution. The effect is to stain brown or black against an almost transparent background various elements of the nervous system and to make clear the details of these elements in a way that had been only incompletely achieved up to that time. Using this method GOLGI was able to demonstrate two particular aspects of certain nervous cells: what he interpreted as an external reticulum, which he had seen before, and what he interpreted by contrast as an internal reticulum, which he had not seen before. In the hands of other investigators, refinements of this method, some of them still in use, yielded additional information about the nervous system generally. Concern here is with the understanding gained about the internal reticular apparatus.

Bothered by comments of other investigators that the impregnation method he had developed was capricious and did not yield comparable results with different types of cells, GOLGI developed, ten years after his initial work on the endocellular apparatus, an arsenious method (1908) for its easier and more consistent demonstration. Using both methods he showed the organelle to exist, although in different form, in the cells of many different types of tissues. He followed its changes in morphology with modifications of cellular activity and its movement within the cell paralleling continued development.

GOLGI himself and most of his students adhered to a belief that they were creating a morphological record and should not speculate on the functions of the organelle. However, they were convinced that they had discovered a new cellular component and were disturbed that it was not generally recognized as such. Dating almost from the time of GOLGI's original paper, cytologists could be divided into nonbelievers and believers in regard to the apparatus' existence. It was not until the advent of electron microscopy and its associated techniques that conversion became the order of the day. Even after clear pictures of the Golgi apparatus had been published there were attempts to explain the organelle as some other component of the cytoplasm.

## D. The Golgi Apparatus and Holmgren's Canals

Early in the century a comparison of the Golgi apparatus was made with the *trophospongium* of HOLMGREN (1902), despite the fact that HOLMGREN's trophospongium had a different distribution in the cytoplasm and was continuous with the exterior of the cell (Fig. 4). This proposition acquired some significance when it was accepted by CAJAL (1908) who made a homology of the cavities of canaliculi of the spongioplasm and the internal reticular

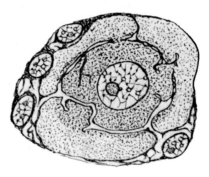

Fig. 4. A cell showing HOLMGREN's trophospongium: a system of canals to which the Golgi apparatus was frequently compared. From Anat. Anz. **20**, 1902.

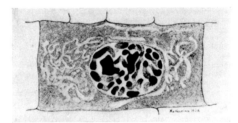

Fig. 5. A system of canals demonstrated by BENSLEY in both animal and plant cells. This diagram is of an onion root tip cell. This contributed to a canalicular interpretation of the nature of the Golgi apparatus. From Biol. Bull. **19**, 1910.

apparatus of cells of the nervous system. This treatment bothered GOLGI (1908) who maintained that a glance at proper preparations would readily distinguish between his internal reticular apparatus and HOLMGREN's trophospongium.

In his later work CAJAL (1914) dropped his comparison of the internal reticular apparatus and the trophospongium and turned to a study of the experimental reactions of what he then termed the Golgi apparatus. In time, consideration of the trophospongium simply dropped out of the cytologist's interpretations.

However, this supposed comparability between the Golgi apparatus and canalicular systems persisted in other forms. BENSLEY (1910) compared the Golgi apparatus to a canalicular system (Fig. 5) and GUILLIERMOND (1929) related it to cavities in the cytoplasm. GUILLIERMOND's assumption that it was represented by cytoplasmic cavities delayed study of the organelle in plant cells because for a long time GUILLIERMOND's work was accepted as a comprehensive treatment of plant cell cytoplasm, which was supposed to have characteristics differing significantly from those of animal cells.

## E. Equivalence with Other Cytoplasmic Components

GATENBY (1955) cites the work of PARAT as the most significant of the attempts to interpret another cytoplasmic component as the equivalent of the Golgi apparatus. PARAT and PAINLEVÉ (1924 a, b) had attempted to equate the Golgi apparatus to what the French cytologists were then referring to as the *vacuome*. The assumption was that exposed to Golgi techniques globules in the cell ran together to form a reticulum. Later PARAT (1928) modified this idea. He divided the *chondriome* (mitochondria) into two classes—active and inactive—and assumed that the inactive mitochondria were also part of what was demonstrated as the Golgi apparatus.

## F. Interpretations as Artifact

A quite different question was raised about the Golgi apparatus by PALADE and CLAUDE (1949 a, b). Working with whole cells and using a high-speed tissue masher they thought ethanol produced myelin figures which were interpreted as the Golgi apparatus. Ethanol is not necessarily used in Golgi preparations, and in a second paper they claimed that the fixatives were responsible for producing the apparent artifacts. However, they discarded the interpretation long ago, and PALADE and his students have been one of the foremost groups in interpretation of the organelle, and CLAUDE has added details about its formation and functioning.

In a quite different category is what BAKER (1944, 1949, 1953, 1955, 1957, 1963) called "the Golgi controversy". BAKER and a group of his colleagues at Oxford denied the existence of a reticulum and tended to relate the structures observed to a modification of a group of spherical bodies which were stained with Sudan black—a stain which had been introduced as a lipid stain.

In his 1957 paper dealing with the Golgi controversy, BAKER adduced evidence that convinced him that the objects in cells which reduced osmium tetroxide or silver nitrate varied from one cell type to another and could not justifiably be designated under GOLGI's name. He also concluded that there was no evidence that there is a "Golgi substance".

BAKER's views tended to change with continuation of the controversy. He came not to deny the existence of cellular components that would react positively to the Golgi methods but to object that they might include many cellular organelles and that GOLGI's name ought not be attached to a variety of components.

For most investigators the existence of the Golgi apparatus and its comparability in different types of cells were established with the adaptation of electron microscopy to biological materials in the early 1950's. Not until 1963 did BAKER become even a reluctant follower. In retrospect it seems remarkable that at the 1954 symposium of the Royal Microscopical Society, GATENBY (1955) referred to some of the early DALTON and FELIX electron micrographs of the organelle as settling the question of its existence and BAKER presented a paper continuing "the Golgi controversy".

## G. Ramón y Cajal

Among those who could be classified as believers from the earliest days was SANTIAGO RAMÓN Y CAJAL, who did enough to improve GOLGI's methods to enable him to launch the modern studies of the Golgi apparatus in a comprehensive and largely experimental paper in 1914. CAJAL's work and GOLGI's were interrelated and, at times, in conflict.

CAJAL, born May 1, 1852, in Petilla de Aragon, Navarre, was a Spanish physician who was trained at Zaragoza. While still a student he developed an intense interest in certain aspects of biological structure and pathological conditions. He received his degree in 1873. Shortly thereafter CAJAL enlisted in the Spanish army which transferred him to Cuba. Not long after arriving in Cuba he contracted malaria which added great misery to his assignment.

Back in Spain after a relatively short interval he took up teaching and research with enthusiasm. His interest in histological research was the more remarkable because the conditions for research in Spain were poor and interest was not high. The atmosphere surrounding his early efforts seems to have been quite the opposite of that which DA FANO (1926) described as surrounding GOLGI's. Years later CAJAL was to note that his research rarely got beyond the borders of Spain until he invested his personal resources in publication of his work. Nonetheless he became a pioneer student of the nervous system and a notable investigator of a number of other subjects. He published many papers and was honored worldwide for his contributions. Like GOLGI his career extended over a long period, his last scientific publication appearing in 1934. He early became interested in the methods GOLGI used to demonstrate the components of the nervous system. Applying these methods as he refined them, he came to the conclusion that the cells of the system did not make up a continuous network as GOLGI supposed but that each cell ended just short of the next.

In 1906, GOLGI and CAJAL were jointly awarded the Nobel prize for their work on the nervous system. For his acceptance speech, GOLGI resurrected his old interpretation of the nervous system as a network; this incensed CAJAL, who felt the idea long disproven and who immediately sought the support of many eminent cytologists for his own interpretation. This may not have been all that aroused CAJAL's ire; years later after he became world famous he wrote in his *Recuerdos de mi Vida* (1923) a note of the surprise incurred by the award of the prize for peace to THEODORE ROOSEVELT and commented at some length on the reduced level at which the Spanish government supported its emissaries abroad. Nevertheless CAJAL continued to be an admirer of GOLGI, repeatedly calling him the "savant of Pavia". In his later years CAJAL recalled rather wistfully that on his way back from visiting German scientific centers in 1889, he had made a trip to Pavia hoping to meet GOLGI and show him some preparations and engage in a discussion of interpretations. He felt that the "polemics and vexatious misunderstandings" which developed later might have been avoided had GOLGI not been away from Pavia when he stopped by.

## H. Cajal's Major Experiment on the Golgi Apparatus

CAJAL's 1914 investigation of the Golgi apparatus is really the starting point of much of the later work. He accepted the idea that the substance and not the specific form gave the organelle its significance and believed it behaved in reproduction in something of the manner described by PERRONCITO (1910) to take on more complex structure after division in certain types of cells under specific conditions. He contributed to knowledge of the wide distribution of the apparatus by finding he could impregnate it not only in nerve tissue but in numerous other types of cells: those of the parotid, submaxillary, pancreas, adrenal, thymus, thyroid, hypophysis, epiphysis, testicle, ovary, digestive glands, connective tissue, adipose tissue, bone cartilage, odontoblasts, spleen and lymphatic glands, and all elements of the embryo. He found that only the liver resisted his attempts at impregnation. (This is curious inasmuch as in modern times liver cells have become one of the favored objects for the study of the apparatus.) Like GOLGI he discovered it would impregnate more readily in young tissues. CAJAL's work was, however, more than simply observation. By various means he directed attention to questions of function and to experimental modification.

The relation of the Golgi apparatus to the process of secretion had been part of the thinking of some investigators since the apparatus was first described. As early as 1902 FUCHS observed that this might be one of its roles, and there is repeated mention of the possibility until it was finally established as a fact and borne out by the work of NASSONOV (1923, 1924) and BOWEN (see 1929).

Referring to the work of several investigators including GOLGI himself (1909), CAJAL made note of numerous changes in the position and form of the apparatus during development and secretory cycles. He made specific reference to the work of WEIGL (1912) and that of DUESBERG (1912) as showing marked changes in stainability and appearance during secretion but joined these investigators in holding that this should not be taken to mean a direct involvement of the apparatus in secretory function. He made reference to regressive metamorphosis of the Golgi apparatus in adipose cells and to its hypertrophy during the secretory phase of odontoblasts and osteoblasts.

He touched directly on the alternation of quiescent phases and progressive phases in calciform cells in mucin secretion as observed by HEIDENHAIN (1907) and MÖLLENDORFF (1913) and he also referred to the work of TELLO (1913) who recognized morphological variations related to secretory function in human mammary gland cells. He commented on the work of NEGRI (1900), reporting on the functioning of the Golgi apparatus in the parotid, the thyroid, and the pancreas of the cat and noted that he (CAJAL) had repeated these observations with different techniques. It is surprising that CAJAL could note as many changes in the Golgi apparatus associated with secretion as he did without mentioning its general role in the process. However, he followed a rigid rule on this score and associated the changes in the apparatus only with variations in the metabolism of the cell. He did go so far as to suggest that the strong secretory excitation provoked by pilocarpine in salivary glands caused a reduction of materials stainable by silver precipitate which might be produced by the Golgi apparatus.

A considerable part of Cajal's paper is taken up with the form of the Golgi apparatus in certain types of nerve cells in relation to the characteristics of these cells. He proposes tentatively that the apparatus encloses material which is consumed during periods of activity and accumulated in quiescent phases. His work on the relation of the Golgi apparatus to traumatic modification of the nervous system is less clear-cut; sometimes the apparatus is appreciably modified; at other times it appears to resist modification. Such experimental work is notoriously difficult to control and interpret. He observed that in the case of cell death the Golgi apparatus is particularly susceptible to autolytic breakdown.

The concluding portion of Cajal's paper is about the normal Golgi apparatus. After a discussion of its location in various types of nerve cells, he proceeds to a separation of the Golgi apparatus and the portion of the chondriome to which it had been compared. He agrees with Duesberg (1914) that a relationship exists between the Golgi apparatus and the attraction sphere (which contains the centrioles), an observation made by many investigators. Finally, he returns to the homology between the Golgi apparatus and Holmgren's trophospongium.

Cajal's paper stands out from the preceding papers in being a methodical treatment that calls attention to the existence of the apparatus in diverse types of cells and to its variation from one cell type to another and from one phase of activity to another. He verified Golgi's contention that this was indeed a newly found organelle; he followed its ontogeny and described its changes with differentiation of the cell; he recorded other changes apparently related to cyclic phases of metabolism in the cell; he introduced chemical agents modifying cellular metabolism and noted that they also modified the Golgi apparatus; he reviewed the breakdown of the organelle and commented on its extreme vulnerability at cell death; and finally, he came very close to concluding that the organelle was involved in secretion.

## III. Secretion

### A. The Secretory Activity of the Golgi Apparatus

Bowen (1929) gives most of the credit for relating the Golgi apparatus to secretion to Nassonov and specifically notes the inaccuracy of crediting this concept to Negri (1900) which was done by several investigators. Interest in secretion actually predated the cell theory. When it was established that secretion was a cellular function (see Bowen 1929), attention turned to the question of what cellular components were involved in the formation of secretory materials. (Secretion, as defined by BOWEN, had a special meaning. See below.) From the discovery of the Golgi apparatus there were repeated contentions that this organelle might somehow be involved. Not until NASSONOV (1923, 1924) pointed out the consistent association of secretory products and the Golgi apparatus and noted certain common staining reactions was the connection made clear. NASSONOV wrote that the papers available to him (limited by the scant penetration of scientific papers from Western Europe into Russia during the Revolution) could be divided into three categories: the first included papers which categorically denied the idea of participation of the organelle in secretion. In this category he included NEGRI's 1900 paper. BOWEN (1929) has remarked that many investigators assumed this paper to establish an association between secretion and the Golgi apparatus but that this assumption is not supported by a careful reading. In certain secreting cells NEGRI failed to record any change in the Golgi apparatus accompanying secretion.

In the second category of papers, NASSONOV put those which indicated only a change in the form and location of the apparatus in relation to the accumulation of secretory material. Among these he included investigations by MARENGHI (1903) on the epidermis of the Ammocoetes; VON BERGEN (1904) on the pancreas; GOLGI (1909) on the epithelial lining of the stomach of the frog; KOLSTER (1913) on the pancreas and some other glands; and CAJAL (1914) who had shown various changes in the apparatus. In his third category NASSONOV listed a group of papers including ones by FUCHS (1902), BIONDI (1911), DEINEKA (1916), and KOLATSCHEV (1916) which actually associated the activity of the apparatus with the function of secretion. BIONDI described the changes in the apparatus with the functional states of cells which secrete cerebrospinal fluid. DEINEKA explored changes in the apparatus in cartilage cells and the development of bony structure whereas KOLATSCHEV studied the cells involved in the formation of glycogen, and lipochrome in nerve cells of molluscs. Most of these papers relate the functioning of the Golgi apparatus during secretion to the chondriome. NASSONOV made an important contribution when he dissociated the activity of the apparatus in secretion from the chondriome. He noted that closest to his own observations were those of FUCHS (1902), who investigated the secreting cells of the epididymal epithelium of the mouse. FUCHS reported that he could follow the formation of granules or drops from the coils of thread making up the apparatus, and he also noted that this aspect of accumulation of granules or drops and their discharge was cyclic. Even FUCHS, however, went so far as to point out that he did not wish to make a categorical statement concerning the secretion of the material, and he concluded that the Golgi apparatus might be an agent acting between the formation of the secretory product and its final discharge. According to NASSONOV, only DEINEKA and KOLATSCHEV suggested that there is an immediate conversion of some regions of the Golgi apparatus into secretory material. He considered that his own results compared in one detail or another to those of FUCHS, DEINEKA, and KOLATSCHEV.

NASSONOV studied a number of glands producing different types of secretory products. He observed the same relationships between the Golgi apparatus and accumulating secretory materials in all of them and it was on some of these glands that study became concentrated after BOWEN (see 1929) focused on the role of the Golgi apparatus in secretion partly by calling attention to NASSONOV's findings.

NASSONOV came to a series of clear-cut conclusions. The first was that in all tissues studied, the primary granules, as well as the minutest mucous drops, appear initially in the meshes of the Golgi apparatus and that it was sometimes possible to see these primary granules in the form of "string-of-pearl-like" arrangements. He contended that as a second step, the granules, having reached a certain size, break loose from the apparatus and collect near the luminal surface of the cell. In no instances did he observe any disintegration of the structure though he did note that the granules that have broken away from the apparatus might at some times fill the whole cell body except for the immediate zone between the nucleus and the adjacent Golgi apparatus. He made a rather fine distinction between production of these materials by the Golgi apparatus and their exteriorization by the cell and concluded that the latter did not depend immediately upon the activities of the organelle. This distinction borne out by modern investigations is responsible for basic differences in Golgi apparatus function and its early elucidation ought not be overlooked as it is a matter of considerable importance.

NASSONOV also noted some differences between the mucin granules and those from the pancreas and the pelvic glands, thereby establishing the basis for chemical differentiation in the organelle. NASSONOV's contributions established the association between the apparatus and secretion, separated secretion and transport, and introduced concern for the differentiation of the organelle.

## B. Bowen's Review

BOWEN's "The Cytology of Glandular Secretion" (1929) is a masterly review of the historical theories of secretion and an acute conception of the components involved. In dealing with the cytological aspects of secretion, BOWEN considered in turn the various hypotheses that had been set forth regarding what component or components of the cell might be involved in the process. His paper culminates in the conclusion that the Golgi apparatus represents a focal point in secretion. The paper is, however, by no means strictly a review paper. It was preceded by a number of investigations by BOWEN himself partly on the role of the organelle in the formation of the acrosome (for references, see BOWEN's bibliography), and partly on the functioning of various glandular cells. Only passing notice will be given here to either the historical work or BOWEN's investigations on the formation of the acrosome. Principal attention will be on BOWEN's interpretations of the Golgi apparatus and secretion from various glandular cells.

BOWEN looked upon secretion as a cytological process by which materials were built up in the Golgi apparatus and then separated from it by a process which differed from one cell type to another. These materials were then "excreted" to be replaced by new secretory products through the activity of the apparatus which remained more-or-less intact. BOWEN recognized the difficulty of separating events by terminology, and it was clear to him that by the time he wrote, the word "excretion" had come to have other meanings. He discussed alternatives of terminology but without enthusiasm for the possibility of substitution. The point of importance is that he recognized the formation of the secretory product as involving intracellular events and looked upon the exteriorization of this product as controlled by other factors.

It is of significance that in his examination of ideas concerning secretion, he touched upon, in an orderly sequence, each of the organelles then detectable that are now known to have a role in the process of assembling most secretory material.

If CAJAL's 1914 paper inaugurated the modern period of study of the Golgi apparatus, it was BOWEN's 1929 paper that gave this study a specific direction, and it began to broaden recognition of secretion as a cellular function. BOWEN made a significant contribution when he pointed out that "apparently the Golgi apparatus plays some immediate role in the process of accumulation and final synthesis of the secretion-products, but the concomitant changes in other cellular structures suggest that all parts of the cell contribute in some way". BOWEN's work has been interpreted by some investigators (DALTON 1961) as drawing some attention away from NASSO-NOV's conclusion concerning the relationship between the Golgi apparatus and the formation of secretory products and focusing it on the relationship of the apparatus to the acrosome. Actually, BOWEN saw a comparable process here, and his only significant qualification was to point out that at the time it was not known how the Golgi apparatus works in the final formation of the secretory product. In many instances it is still not known, and DALTON (1961) is correct in his statement that many new discoveries have served only to complicate the problem.

Before turning to other considerations about the Golgi apparatus it seems in order to remark that the light microscopists made great strides in studying an organelle not altogether consistent in reactions to techniques, extremely variable from one cell type to another, and apparently in different phases of activity. To complicate matters the initial demonstration had been of a form that is somewhat less typical than some others. As DALTON (1961) has pointed out, it took more adequate instrumentation to derive a clear concept of the organelle in its different forms, and that stage was not reached until it became possible to apply electron microscopy to study of the apparatus in various cell types. Apparently, the first micrographs showing the recognizable components of the Golgi apparatus were published by DALTON and FELIX in 1954. These micrographs were in press when the Royal Microscopical Society held its symposium on the Golgi apparatus in 1954, but DALTON and FELIX kindly furnished copies of their micrographs to GATENBY for his review of the Golgi apparatus at the symposium (GATENBY 1955). It is significant that the early electron micrographs of the Golgi apparatus depicted the form which BOWEN had come to recognize as a common one and the importance of BOWEN's investigations and review is revealed in the concentration of study on the relationship between the organelle and the formation of secretory products.

# IV. The Golgi Apparatus in Plant Cells

During the early investigations there was little exploration of the possibility of the presence of the Golgi apparatus in plant cells. Several factors contributed to this situation. The early presumed analogy between the Golgi apparatus and the canals of HOLMGREN (1902) led to an assumption of equivalence between the Golgi apparatus and elements of the vacuome. This idea was supported in one form or another by BENSLEY (1910), GUILLIERMOND and MANGENOT (1922), and GUILLIERMOND (1929). The association of the apparatus with secretion tended to de-emphasize plant cells, only a few types of which were recognized as secreting cells. Nonetheless, BOWEN (1926) concerned himself with the question of whether plant cells contained Golgi apparatus. He developed an idea that the plastids might equate to the Golgi apparatus. This idea was picked up by WEIER (1932) who gave credit for it to BOWEN. BOWEN subsequently undertook a detailed study of plant cells and concluded that he could demonstrate the normal components of animal cells and something in addition, to which he applied the term *osmiophilic platelets* which were much more demonstrable in certain tissues than in others (BOWEN 1928). Having done so, he accepted the equivalence of these osmiophilic platelets and the Golgi apparatus.

BEAMS and KING (1935 a, b) performed a rather critical experiment when they demonstrated that the Golgi apparatus of animal cells and the osmiophilic platelets of bean root tip cells are displaced to the same level by ultracentrifugation. There were thus elements in plant cells which, impregnated with osmium, had the same structure as some Golgi apparatus and showed the same specific gravity.

When KIRKMAN and SEVERINGHAUS wrote their three-part review of the Golgi apparatus (1938 a, b, c) they simply assumed the presence of the Golgi apparatus in plant cells without too much concern for which homology might explain the proposition. BEAMS and KESSEL (1968) pointed out that it did not become respectable to talk about the Golgi apparatus as an organelle of plant cells until the early 1960's after a number of studies by electron microscopy had been made. As WHALEY et al. (1959) indicated, "A Golgi-apparatus (comparable to that of animal cells) has been demonstrated in cells of *Allium* by PORTER (1957); of *Allium, Triticum,* and *Chrysanthemum* by BUVAT (1957 a, b); of *Zea, Vicia, Aneura,* and *Antirrhinum* by HEITZ (1957 a, b, c); of *Zea, Trianea, Cucurbita, Vicia, and Chlorophytum* by PERNER (1957, 1958); *Triticum* by SETTERFIELD and BAYLEY (1958); of *Pisum* by SITTE (1958); of *Nitella* by HODGE et al. (1956); of *Elodea* by BUVAT (1958); of several cryptogams by HEITZ (1958); and of various algae by DALTON and FELIX (1957), ROUILLER and FAURÉ-FREMIET (1958), CHARDAR and ROUILLER (1957), and SAGER and PALADE (1957)". Despite the length of this list, many of the earlier studies did not provide sufficient information to allow one to draw unequivocal conclusions concerning the organelle or its relation to the endoplasmic reticulum. As a matter of fact it took many progressive changes in the techniques and preparative procedures to make clear the character of the Golgi apparatus in plant cells. This is hardly surprising in view of the fact that many useful procedures in animal cytology have proved less than ideal for application to plant cytology. When techniques had reached a certain point it became clear that the Golgi apparatus is a normal organelle of all eukaryotic plants at certain stages. The establishment of the existence of the Golgi apparatus in plant cells makes a reconsideration of its role in general cellular activities mandatory. Its recognized function in secretion suggests that plant cells not previously classified as secreting should be examined for evidence of secretory functioning. But beyond this, the possibility of an even more fundamental role for this organelle needs exploration.

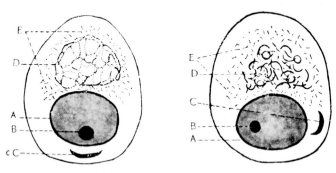

Fig. 6. PERRONCITO's original diagram illustrating dictyokinesis and the elements to which he applied the term *dictyosome*. Left. A cell before dictyokinesis begins. $A$ = nucleus, $B$ = nucleolus, $C$ = centrosome? $D$ = Golgi apparatus, $E$ = chondrosomes of Meves (mitochondria). Right. Dictyokinesis in progress. $A$ = nucleus, $B$ = nucleolus, $C$ = centrosome? $D$ = dictyosomes, $E$ = chondrosomes of Meves (mitochondria). From Arch. ital. Biol. **54,** 1910.

## V. Form

### A. The Dictyosome

When PERRONCITO (1910) studied the activities of the Golgi apparatus during cell division he recorded the fact that it divided into smaller units which sometimes had definite patterns of distribution to the daughter cells. To each of these smaller units he applied the term *dictyosome* and the term *dictyokinesis* to the total process of division and distribution (Fig. 6).

BOWEN in his extended studies observed that in many types of cells the Golgi apparatus exists in lamellar form. It had been observed by numerous investigators, as for example HIRSCHLER (1918) in invertebrates, that this group of organisms is characterized by the lamellar form. The term dictyosome was applied to this form to indicate the difference between it and the reticulate form which GOLGI (1898) first demonstrated. When the apparatus was unequivocally demonstrated in plant cells it was shown to be in this dictyosome form. Over time, the dictyosome form has proved to be of much more common occurrence than the reticulate form, though the problem of distinguishing between the two is complicated by interconnections and the contiguousness of the dictyosomes in many instances.

The term Golgi apparatus will be used here in preference to dictyosome because 1. electron microscopy demonstrates fairly discrete groups of lamellar units, 2. PERRONCITO had earlier established a definitive use for dictyosome, and 3. each dictyosome appears to function in membrane building, product synthesis and assembly, and in the evolution of products as an individual

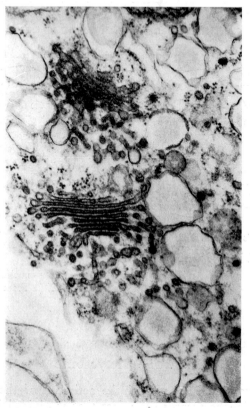

Fig. 7. A portion of a cell of *Nitella* showing a typical stack of Golgi cisternae and, in a different view, variations in the form of these cisternae. Also showing different types of vesicles associated with the stack. From F. R. TURNER, Cell Research Institute, University of Texas at Austin. ×28,000.

organelle. It must be recognized, however, that dictyosome is firmly entrenched in the literature and is preferred by many investigators for parts of or for the entire Golgi apparatus. MOLLENHAUER and MORRÉ (1966) contend that the dictyosomes in the cell are interconnected by membranous elements of the system itself to constitute the Golgi apparatus (see also RAMBOURG *et al.* 1973). NOVIKOFF *et al.* (1971) have evidence suggesting interconnections by lamellar elements of the endoplasmic reticulum. The problem is further complicated by the fact that if one views impregnated cells in light microscopy one frequently sees a reticulate system whereas the same cells show more or less discontinuous stacks of lamellae in electron microscopy.

It now appears that in one sense the dictyosome can be looked upon as a relatively simple form of the organelle. It characterizes the egg and certain embryonic tissues. In this *dispersed* form, as WILSON (1925) referred to it,

Fig. 8. A freeze-etched section through the Golgi apparatus of an alga, *Micrasterias denticulata,* showing an arrangement of cisternae and associated vesicles comparable to that seen in plastic-embedded sections. From STAEHELIN and KIERMAYER, J. Cell Sci. **7**, 1970. Courtesy of Cambridge University Press. ×52,000.

it characterizes stages of cell division generally. In certain cells the onset of some activities is characterized by a localized Golgi apparatus resulting from either extensive growth or aggregation of dictyosomes. In other cells differentiation does not seem to be accompanied by the development of further complexity in the apparatus.

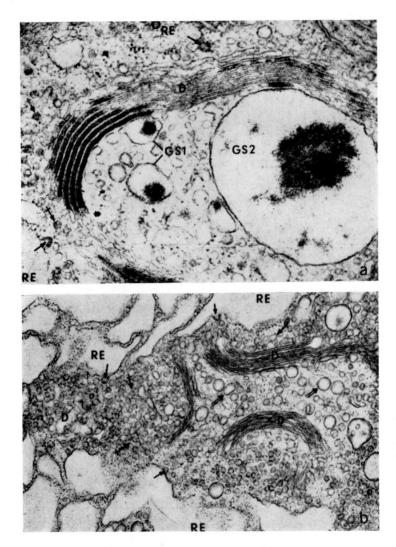

Fig. 9. Modification of the Golgi apparatus in the multifid gland of a snail in different stages of the annual cycle of activity. *a* Cell in secretory activity. *b* Cell at rest. Simple arrows indicate evaginations from the endoplasmic reticulum which may become transition vesicles (barred arrows). Double arrows indicate a different class of vesicles that arise during inactivity. *D* = dictyosome, *GS* = secretory granules in different stages of maturation, *RE* = endoplasmic reticulum. From OVTRACHT, J. Microscopie **15**, 1972. *a*) ×38,000, *b*) ×25,000.

## B. The Golgi Stack

Instead of referring to the central structure of the Golgi apparatus as a dictyosome, it will simply be called a *stack*. This stack is made up of flattened membrane-bounded cisternae with which are associated various vesicles (Fig. 7). It appears in the same form in unstained freeze-etch material as it does in conventionally fixed and embedded material (Fig. 8) (see BRANTON and MOOR 1964, STAEHELIN and KIERMAYER 1970). It is in the stacked form that the organelle has been mainly studied. The terms *cisterna* and *saccule* have been variously and interchangeably used in describing the

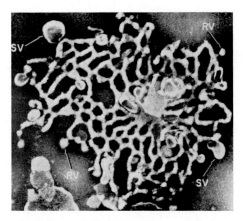

Fig. 10. An early micrograph of an isolated cisterna from *Allium cepa* much modified by isolation procedure and phosphotungstic acid staining. $SV$ = smooth vesicles, $RV$ = rough vesicles. From CUNNINGHAM *et al.*, J. Cell Biol. **28**, 1966. Courtesy of Rockefeller University Press. ×38,000.

individual membrane-bounded components of the stack. Cisterna will be used here.

There are apparently a number of variations of this form. In some secreting cells such as the goblet cells (NEUTRA and LEBLOND 1966 a) cisternae having reached a certain stage of maturity appear to be entirely transformed into mucigen granules which essentially fill the goblet of the cell (Fig. 46). In a few instances (OVTRACHT 1972) certain stages of development are characterized by a partial breaking up of the stack into small vesicles which are apparently reorganized somehow into the stacklike form when the organism prepares for re-entry into the secreting phase (Fig. 9). The morphology of the organelle is greatly modified by certain experimental techniques and notably by attempts to isolate it from the cell (Fig. 10) (CUNNINGHAM *et al.* 1966, MORRÉ *et al.* 1970). Considerations of the form of the apparatus also enter into the question of whether certain secretions characteristically flow through it to acquire segments of membranes and into questions concerning its replication. HAY and DODSON (1973), considering the question of secretion of intercellular matrix material, have raised the possibility that the Golgi apparatus may exist in some instances in a hitherto unrecognized form. The failure to prove a direct means of replication has raised the possibility that it may alter its form and replicate in a stage not commonly recognized (see Section XI).

The number of Golgi apparatus per cell is apparently a feature of differentiation. Some cells have been recorded as having only a single apparatus

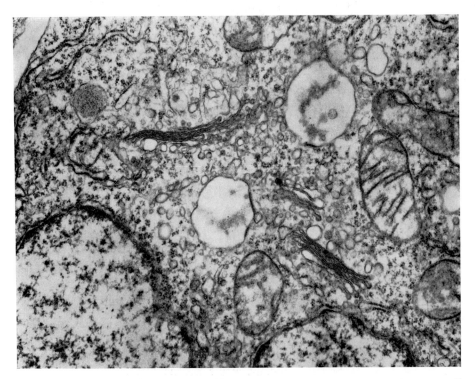

Fig. 11. Portion of a cell from a late blastula stage of an unknown species of water snail showing dispersed Golgi apparatus. From DAUWALDER *et al.*, Sub-Cell. Biochem. 1, 1972. ×23,000.

relatively minor in extent. Other cells with dispersed Golgi apparatus may have hundreds. In some of the classical glandular secreting cells where the apparatus is localized there appears to be only a single Golgi apparatus but of appreciable extent. In such cases the organelle may occupy a rather specific position (Fig. 46), whereas in the instances of dispersed Golgi apparatus there is a tendency for the organelles to occur generally throughout the cells (Fig. 11). Large amounts of secretory product may be produced by intensely active single apparatus or they may be a result of coordinated activities of many apparatus. They are in any event correlated in some instances with production of large amounts of membrane, apparently sometimes in a relation to cell growth, in others only with the rapid exocytosis of the material.

The number of membrane-bounded cisternae in a stack differs appreciably from species to species, sometimes from one stage of development to another, or from one stage of activity to another. There may be no more than 5 or 6; or as many as 25 or 30 (compare Figs. 7 and 12). Whether there is sometimes only a single cisterna raises problems that have not been answered regarding the origin of the Golgi apparatus and the manner of its replication (see Section XI). The number of cisternae and their activity in the formation of

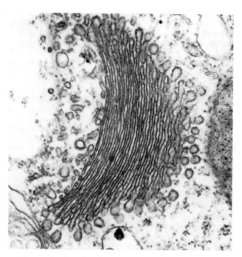

Fig. 12. Golgi apparatus of an individually discrete type from *Euglena* showing a large number of cisternae. From ARNOTT in WHALEY, in: The biological basis of medicine, Vol. 1 (BITTAR, E. E., and N. BITTAR, eds.). London-New York: Academic Press. 1968. ×30,000.

Fig. 13. A freeze-etched preparation showing peripheral fenestration in a single Golgi cisterna of *Micrasterias*. From BRANTON and MOOR, J. Ultrastruct. Res. **11**, 1964. ×52,000.

different products is a variable not only in terms of the changing activity of the cell but also in respect to experimental treatments (see Section XII).

The Golgi stack, particularly in the simpler, more discrete form, has often been compared to a stack of pancakes. Like many analogies, this one is inadequate and misleading. The pancakes in a stack are usually of fairly

uniform size. The cisternae of a Golgi stack may vary greatly in size, as
indicated in Fig. 12. Often the individual components in the stack are
reticulate structures and not simply flattened cisternae. GOLGI himself recognized the existence of what he termed "little discs" and several earlier
investigators emphasized that they might be fenestrated. The fenestration
may occur only around the periphery (BRANTON and MOOR 1964) (Fig. 13)

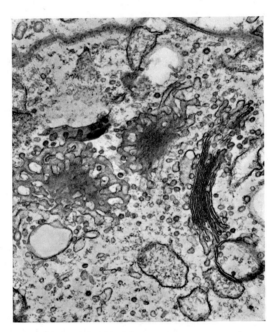

Fig. 14. Section of a cell of a fresh water clam gill showing Golgi apparatus in different
views. The fenestration is very extensive; some of the cisternae are in effect tubulo-
vesiculate structures. From TURNER in WHALEY, in: Organisation der Zelle. III. Probleme
der biologischen Reduplikation (SITTE, P., ed.). Berlin-Heidelberg-New York: Springer.
1966. ×17,000.

or it may characterize three-fourths or more of the entire cisternae (KESSEL
and BEAMS 1965, MOLLENHAUER and MORRÉ 1966, HOFFER 1971) (Fig. 14).
In the instances in which the reticulation is extensive it usually does not
include the centermost portion of the cisternae. The reticulation of the
cisternae may differ in extent from one face of the apparatus to the other,
and it may differ developmentally, being absent or nearly absent at certain
stages and extensive at others (Fig. 15). There is some evidence that the
extent of apparent reticulation may be enhanced both by standard methods
used in preparing materials for study by electron microscopy and procedures
used to isolate the Golgi apparatus. CUNNINGHAM (1974) has studied the
progressive changes in the vesiculation of cisternae associated with various
treatments used to isolate the apparatus for study and has related particular
modifications to particular procedures (Fig. 16). These modifications are
not surprising. Red blood cells, *in vivo* characteristically ovoid, appear after

certain preparative methods as extremely reticulo-vesiculate structures. This modification may be a fairly common reaction of membrane-bounded components to certain methods of preparation. The existence of the reticulation of the Golgi apparatus *in vivo* is, however, attested to by both the evidence from freeze-etch materials and the differences seen in developmental sequences. In some instances what is referred to here as reticulation is so extensive that

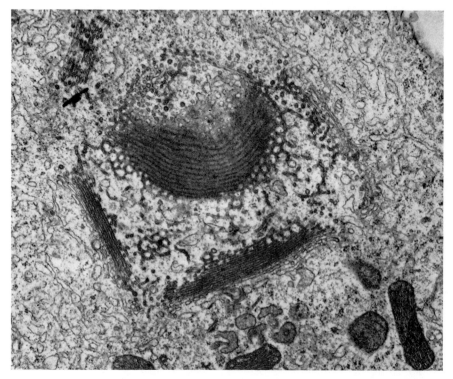

Fig. 15. Golgi apparatus in different views in the spermatid of a snail *Helix aspersa* showing different degrees of fenestration on different faces. From DAUWALDER *et al.*, Sub-Cell. Biochem. **1**, 1972. ×7,600.

the appearance is one of a network of interconnecting tubules. The significance of this reticulation is not known.

FRANKE and SCHEER (1972) have interpreted several instances of reticulation in a distinctive manner. They have proposed that there are some similarities between what they term dictyosomal pores and the pores of the nuclear envelope and annulate lamellae. Such would not be surprising in view of the relationship that has been postulated between the nuclear envelope and both the Golgi apparatus and the annulate lamellae (see LONGO and ANDERSON 1969, KESSEL 1968, 1971, 1973, CHRÉTIEN 1972). FRANKE and SCHEER found the openings in the Golgi cisternae to surround dense particles or filaments which they suggest may be ribonucleoprotein (Fig. 17). They call attention to the existence of similar granules or rods in pores in other cytomembrane

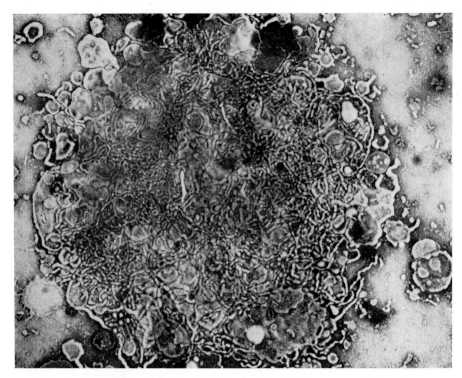

Fig. 16. Isolated rat testis Golgi apparatus stained with phosphotungstate. The stack structure appears much modified in morphology. From CUNNINGHAM, in: Sub-cellular particles, structures, and organelles (LASKIN, A. I., and J. A. LAST, eds.). New York: Marcel Dekker. 1974. ×37,000.

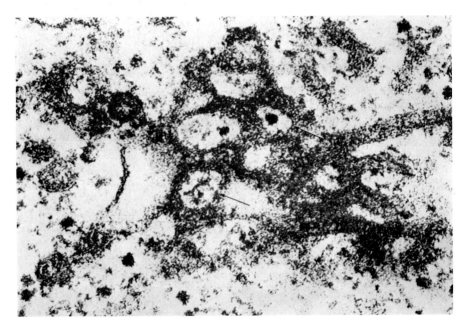

Fig. 17. Cisternal pores showing contents including filaments or particles (arrows) interpreted by the investigators to be ribonucleoprotein. Portion of a tissue culture cell of *Haplopappus*. From FRANKE and SCHEER, J. Ultrastruct. Res. **40**, 1972. ×160,000.

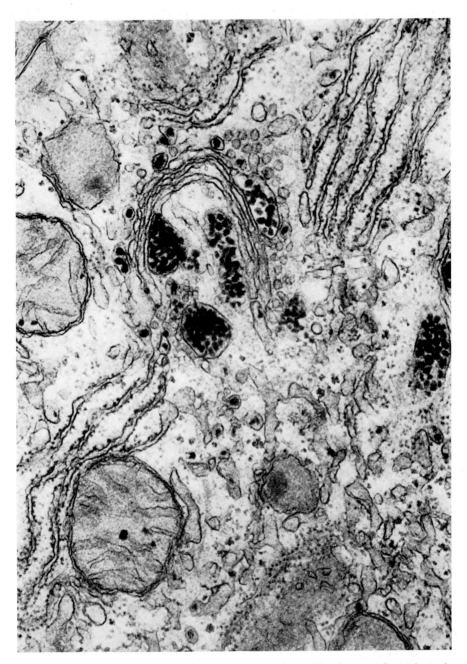

Fig. 18. CLAUDE's interpretation of the formation of the Golgi apparatus from the endoplasmic reticulum in rat liver. The interpretation is based on the sites of the low density
lipoprotein granules which appear dark in the micrograph. From CLAUDE, J. Cell Biol. **47**,
1970. Courtesy of Rockefeller University Press. ×52,000.

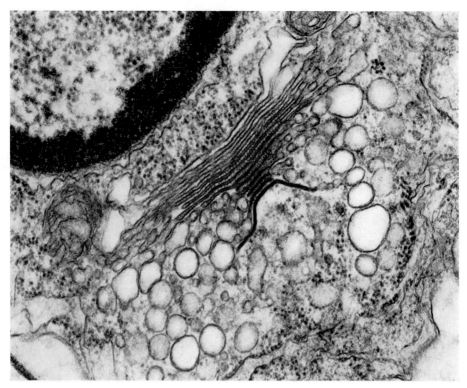

Fig. 19. Golgi apparatus of *Nitella* cell showing intercisternal elements cut perpendicular to their long axes. Note also changes in the density of cisternal contents from one face to the other and apparent evolution of secretory vesicles from cisternae not at the distal face. From Turner in Whaley, in: The biological basis of medicine, Vol. **1** (Bittar, E. E., and N. Bittar, eds.). London-New York: Academic Press. 1968. ×48,000.

systems, citing a number of specific instances, and they hypothesize that the ribonucleoprotein may play some part in the formation of pores rather than simply occupying them.

Claude (1970) has proposed from a study of low-density lipoprotein granules in the endoplasmic reticulum and the Golgi apparatus that fusion of segments of smooth endoplasmic reticulum is responsible for formation of fenestrated sections of the Golgi apparatus cisternae and that continued modifications determine some of the structural differences between the opposing faces of the apparatus (Fig. 18). Whether there is direct transformation of membranes from one system to the other is not clear. There is evidence for the transfer of membrane components from the endoplasmic reticulum to the Golgi apparatus (Whaley *et al.* 1971) and there are functional inter-relationships (see below).

In a few cases elongate elements, 60–80 Å in diameter, have been demonstrated between the cisternae (Turner and Whaley 1965, Mollenhauer 1965 a). When demonstrable, these elongate elements are oriented in the same direction and give the appearance of being in register (Fig. 19). If they can

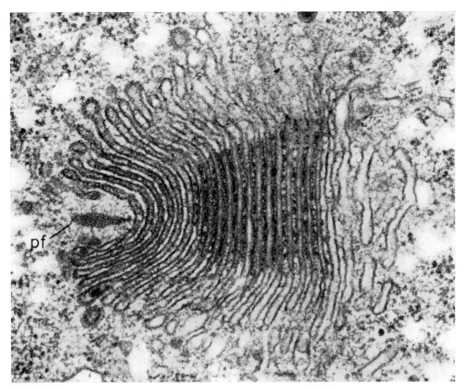

Fig. 20. Golgi apparatus (parabasal body) of *Trichomonas termopsidis*. Dense nodes of intercisternal material do not involve the 5 to 6 proximal spaces or the most distal intercisternal spaces. Transverse section, *(pf)* parabasal filament. From Amos and Grimstone, J. Cell Biol. **38**, 1968. Courtesy of Rockefeller University Press. ×69,000.

be seen only across part of the Golgi stack they are seen toward the distal face. Both the composition and function of these intercisternal elements are unknown. The only certain evidence about them is that they acquire distinctive electron density in glutaraldehyde-fixed, osmium post-fixed preparations. Whether the darkly stained portion represents the whole structure or only a central element is not known. Cunningham *et al.* (1966) have suggested that intercisternal elements may extend beyond the edges of the cisternae. This suggestion was based on micrographs of glutaraldehyde-stabilized, isolated, negatively stained Golgi cisternae and its meaning with respect to relationships in an intact cell has not been established. These intercisternal elements do not appear to be the same as the rodlike elements that Franke and his colleagues have demonstrated as running through the pores of intracellular membranes.

Amos and Grimstone (1968) have described a dense material between the cisternae of the Golgi apparatus in *Trichomonas* and some other flagellates. After glutaraldehyde fixation this material appears more or less uniformly granular (Fig. 20). After osmium fixation certain portions of it appear much more dense than others. The material was confined to certain regions of the

Golgi apparatus to which AMOS and GRIMSTONE applied the term *nodes* (Fig. 21). They noted that the central region occupied by this dense material was one in which the Golgi cisternae had relatively uniform thickness and that the spacing of the cisternae was generally consistent here. AMOS and GRIM-STONE concluded that this material was not formed at the same time as the cisternae because it was not found close to the proximal face. MOLLENHAUER

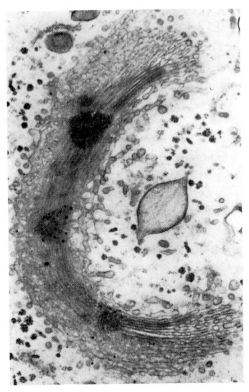

Fig. 21. Oblique longitudinal section of the Golgi apparatus of *Trichomonas termopsidis* showing the dense nodes with small dense granules characteristic of OsO$_4$-fixed material. From AMOS and GRIMSTONE, J. Cell Biol. 38, 1968. Courtesy of Rockefeller University Press. ×25,000.

and MORRÉ (1966) have concluded from the fact that the organelle may maintain its integrity in isolation procedures and the frequent appearance of intercisternal material that the cisternae are held together by an intercisternal cementing substance of some sort. AMOS and GRIMSTONE mention this same possibility, but they raise the question of why intercisternal material is detectable in some instances but not in others. The presence of intercisternal elements or intercisternal material of other sorts makes difficult the interpretation that the Golgi apparatus cisternae are progressively displaced from one face of the apparatus to the other as has been proposed in certain instances (NEUTRA and LEBLOND 1966 a).

Nonetheless AMOS and GRIMSTONE emphasize that in *Trichomonas* the Golgi apparatus is apparently in a steady state with replacement on one side and loss from the other. They then discuss the difficulties this makes in envisioning the formation and insertion of intercisternal material. MOLLEN-HAUER and MORRÉ (1966) have presented a diagram illustrating several of the features discussed here (Fig. 22).

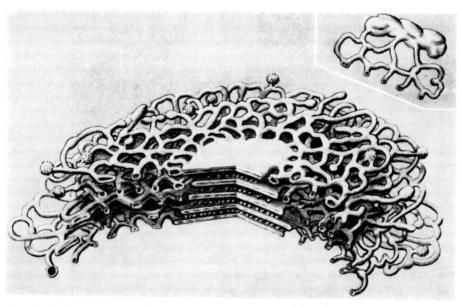

Fig. 22. A diagram of a portion of what MOLLENHAUER and MORRÉ have called the plant dictyosome. They have illustrated an extensive fenestration and the formation of rough-sur-faced vesicles in the cisternae. A different sort of vesicle is shown in the insert. They have also diagrammed intercisternal elements and indicated changes in the cisternae from the top to the bottom of the diagram. Few, if any, Golgi apparatus show all the characteristics diagrammed. From MOLLENHAUER and MORRÉ, Ann. Rev. Plant Physiol. **17**, 1966.

MOLLENHAUER *et al.* (1973) have explored in some detail intercisternal elements and plaques in certain plant Golgi apparatus. They have done this by modifying the environment of the apparatus and in some instances applying negative stains. The implication of their experiment is that the stack of cisternae is held together by the particular composition of the intercisternal materials and by modifying this composition they can separate cisternae in various stages of differentiation. Doing so is a requirement for certain kinds of studies of differentiation within the apparatus, but achieving meaningful results will require more than just simple separation. It will necessitate some method for categorizing cisternae in specific states of development. This is fraught with still further difficulties since it is not known whether the stacks in an individual cell are all in a uniform stage of development.

MOLLENHAUER *et al.* (1968) have called attention to the structure of the centermost region of the Golgi stack of *Euglena gracilis,* pointing out that in this instance there is an accumulation of dense material within the cisternae. The boundaries of the region in which this accumulation is seen are, in general, conical. Their micrographs show, as do those of AMOS and GRIMSTONE, a uniformity in the width of the cisternal lumens in this region, whereas outside it they are distinctly less uniform. Within the region, too, the cisternae appear in parallel orientation.

MANTON (1967 a) has made a related observation showing that in the Golgi stack of *Chrysochromulina chiton* the central region of the cisternae may become locally distended and distinguishable by staining reactions from the more peripheral regions. Her interpretation is that this may represent the accumulation of some unidentified metabolite in the central region which stains densely at certain times of the day. MANTON has tentatively suggested that this particular kind of Golgi apparatus may be a diagnostic feature of the group in which this genus occurs. It is true that tissues and sometimes unicellular organisms may be characterized by relatively distinctive Golgi apparatus.

A most striking demonstration of a dense material between (and possibly also within) the cisternae has been provided by MIGNOT (1965) in a study of the Golgi stack of *Distigma.* The dense core does not appear between the more proximal or the more distal cisternae. It does extend across three-quarters of the stack and is roughly conical in shape being reduced toward the distal face (Fig. 23). MIGNOT interprets the appearance of more than one of these conical regions (which would seem to conform to the nodes described by AMOS and GRIMSTONE) to indicate a stage in the division of the organelle and points to irregularities and continuities in some of the cisternae as suggesting that the Golgi apparatus is actually undergoing division. MOLLEN-HAUER (1971), on the other hand, has suggested that the distalmost cisterna may occasionally leave the stack. Such cisternae could, of course, form focal points for the development of new Golgi apparatus.

Critical information concerning the character of the central region of the Golgi stack is still lacking. The evidence does, however, seem to be adequate to suggest that this region is both structurally and functionally different from the more peripheral regions, and that the differences may be demonstrable by currently available techniques only in certain phases of activity or stages of development.

The most desirable way in which to determine the status of a cellular organelle is to isolate it in a relatively clean fraction and then either provide for a direct analysis of its composition or assay it for enzyme activity in a known substrate. Such combinations of biological and biochemical techniques have been used to advantage with several cellular components. Attempts to isolate the Golgi apparatus and assay it go back to an experiment by KUFF and DALTON (1959). KUFF and DALTON were interested in phosphatase activity of the organelle. BOURNE (see 1955) had cytochemically demonstrated a phosphatase activity in the apparatus and concluded that the activity was hormonally induced.

KUFF and DALTON, using preparations that contained numerous Golgi apparatus, showed the presence of both acid and alkaline phosphatase activity as well as demonstrating substantial amounts of phospholipid and ribo-nuclease. The Golgi apparatus, however, presents some formidable technical difficulties to the usual procedures of isolation. With homogenization its smooth lipoprotein membranes break up to form smooth microsomes. So do the plasma membrane, the smooth endoplasmic reticulum, and some other cellular components. The Golgi apparatus usually has an axis of differentia-

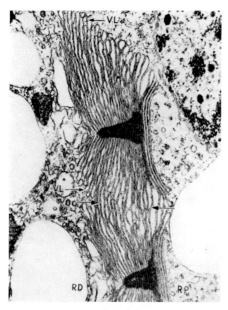

Fig. 23. A micrograph of *Distigma* Golgi apparatus showing material between some of the cisternae as well as perhaps within the cisternae that stains much more densely than the other cisternal contents. The investigator interprets the presence of two cones of material and the anastomosing between as indicating that the Golgi apparatus is in the process of division. *RP* = proximal face, *RD* = distal face, *VL* = associated vesicles. From MIGNOT, Protistologica **1** (2), 1965. ×20,000.

tion from one face to the other involving changes both in membranes and synthesis and assembly of products. As a result, isolated Golgi fractions represent averages of variously differentiated components. Further, membrane-bounded vesicles derived from it *in vivo* are known to undergo substantial change once they have been separated, and these smooth membrane-bounded components may further complicate the analysis of fractionation data.

Nonetheless, the technical difficulties have not prevented students of the organelle from attempting isolations. Different groups of investigators have utilized different procedures. Some of the procedures have involved the use of additives, a few of which like glutaraldehyde seem to have some effectiveness in stabilizing the structure (MORRÉ and MOLLENHAUER 1964, FLEISCHER and FLEISCHER 1971, BOWLES and NORTHCOTE 1972). Interpretations of glutaraldehyde-stabilized preparations must, however, take into account the fact that glutaraldehyde is known to modify some structures as well as fixing them. Other procedures have involved the induction of low-density lipids which have an influence on isolation and serve as markers (EHRENREICH *et al.* 1973). At this stage it is only possible to derive certain fractions that can be designated as Golgi apparatus-rich fractions. Even so, it has been possible to characterize such fractions in certain ways and to correlate the results with data from other procedures. To avoid repetition these data will be discussed

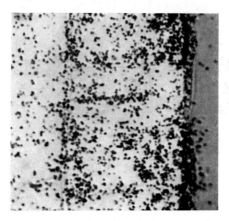

Fig. 24. A light micrograph showing the difference in the polarity of secretion in cells labelled with ³H-glucose in epidermal cells of *Zea mays*. Interphase cells at top and bottom, dividing cell just above center. In the interphase cells there is heavy transfer of label in membrane-bounded vesicles to the surface; in the dividing cell this transfer is reduced and there is heavy transfer to the cell plate. From DAUWALDER *et al.*, J. Cell Sci. **4**, 1969. Courtesy of Cambridge University Press. ×1,000.

in the later parts of the work where they are most pertinent to the interpretations of specific aspects of Golgi apparatus functioning.

Although GOLGI's original recording of the apparatus (1898) was of a reticulate type, many of the early investigations were made on cells recognized to be secreting cells which had distinctive Golgi stacks. Many such cells characteristically have a long, curved Golgi apparatus in a juxtanuclear position on the luminal side of the cell, and there is a progressive maturation of secretory granules between the apparatus and their point of exit from the cell. Thus the entire cell is highly polarized in an anatomical sense and the secretory products move uniformly in a single direction. Such polarized cells give an impression that the direction of secretion is controlled by the position of the Golgi apparatus. In fact, there are some cells such as the follicular cells of the thyroid (see Section VI-D) in which the Golgi apparatus shifts position in relation to the direction of secretion. Studies of cells with dispersed Golgi apparatus have shown, however, that there may be distinct polarity of secretion unrelated to any fixed position of the Golgi apparatus (Fig. 24). The direction of such secretion may change with the particular stage of activity of the cell.

## C. Activities across the Stack

Most studies have suggested a buildup of activities across the Golgi stack. Although the detection of particular activities on opposite faces or even in various parts of cisternae has now made this seem an oversimplification, there may still be a strong buildup of certain activities from one face to the other.

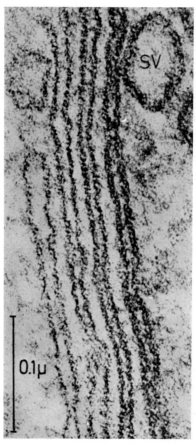

Fig. 25. Changes in the thickness of Golgi apparatus membranes in the fungus *Pythium ultimum*. Those toward the proximal face, at the left, approximate the nuclear envelope and endoplasmic reticulum in thickness; those toward the distal face, at the right, approximate the plasma membrane in thickness. *SV* = secretory vesicle. From GROVE *et al.*, Science **161**, 1968.

This appears to be the case in the goblet cell studied by LEBLOND and his colleagues (PETERSON and LEBLOND 1964 a, b, NEUTRA and LEBLOND 1966 a, b, 1969) and may even relate to some specific patterns of secretion. LEBLOND (personal communication) has pointed out that in different circumstances and in different sorts of cells there may be almost no easily detectable differences between one face of the apparatus and the other or there may be very distinctive differences. Such differences may, for example, appear in membrane thickness (see Fig. 25). They may be visible in hypertrophy of cisternae indicating synthesis and accumulation of secretory products or they may be detectable in lysosomal enzymes as demonstrated by cytochemical staining. The basis of such differences is unknown.

Sometimes the cisternae at one face of the Golgi stack are curled and those of the midregion or toward the other face of the Golgi stack may be flat or nearly so (Fig. 26). A number of investigators have used curling as a means of distinguishing between the two faces of the Golgi stack, often

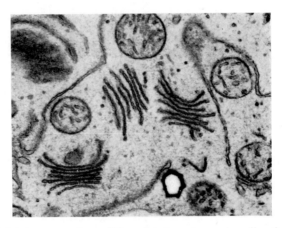

Fig. 26. Simple Golgi apparatus in undifferentiated meristematic cells of *Zea mays* root showing both flat and curved forms and flatness and curvature within the same stack. From M. DAUWALDER, Cell Research Institute, University of Texas at Austin. ×20,000.

referring to one face as convex, the other as concave. In many instances the concave side has been assumed to equate to what has here been described as the distal face and the opposite or convex side to the proximal face. In a few instances, the opposite interpretation has been made. The reliability of curling as such an indicator is uncertain for it is not clear whether it relates to structural factors present in the central core, extension of the cisternae, activities within the stack, or whether it is a consequence of motion within the cytoplasm generally, or a modification resulting from fixation. In some Golgi stacks the cisternae on the two faces may be curled in opposite directions.

In Golgi apparatus obviously involved in the processing of secretory material the commonest pattern appears to be a continuing buildup of such material toward the distal face. This accumulation may be visible in terms of increased lumen width (Fig. 27) or localized distensions and often by distinct staining reactions. The buildup may involve accumulation, transformation, synthesis, and conjugation. There are too many variations in Golgi apparatus activity and too few data about the actual functioning of the organelle to generalize but two points can be made about the secretory product: the enzyme, hormone, or other active principle involved may be variously associated with other products that modify its activity or most likely make it inactive while it is in the cytoplasm; and the rate of differentiation for secretion differs among cell types as does the intensity and the length of the period during which the cell is actively secretory.

Depending upon the cell type and perhaps on the relation of the secretory product to the membrane, patterns of synthesis and assembly of this product indicate many different forms of activity on the part of the Golgi apparatus. The product may be accumulated within the cisternae and then pinched off in membrane-bounded form. Usually more product is seen toward the distal

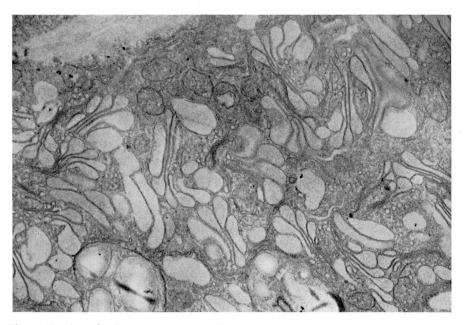

Fig. 27. Portion of a *Zea mays* rootcap cell showing hypertrophy of some Golgi apparatus which extends from the distal to the proximal face. From H. H. MOLLENHAUER, Cell Research Institute, University of Texas at Austin. ×15,000.

face but in some instances it may extend to the proximal face (Fig. 31) as well. In either case some product may be seen within the central portion of the cisternae, but it apparently moves to the periphery before being pinched off. In the case of the pancreatic exocrine cell the secretory product appears to be accumulated and modified in the so-called condensing vacuoles near the distal face of the apparatus, the membranes of which are presumably derived from the Golgi stack (Fig. 39). In such cases secretory material may occur only at the periphery of individual cisternae to which they are apparently moved before being pinched off.

The increasing evidence (see CAPALDI 1974) of significant differences in membrane character plus the general evidence that membranes are reaction surfaces may lay a basis for explaining some of the activity differences in the light of structural differences in the molecules of the membranes or molecules closely associated with them. If the emphasis is put primarily upon membrane

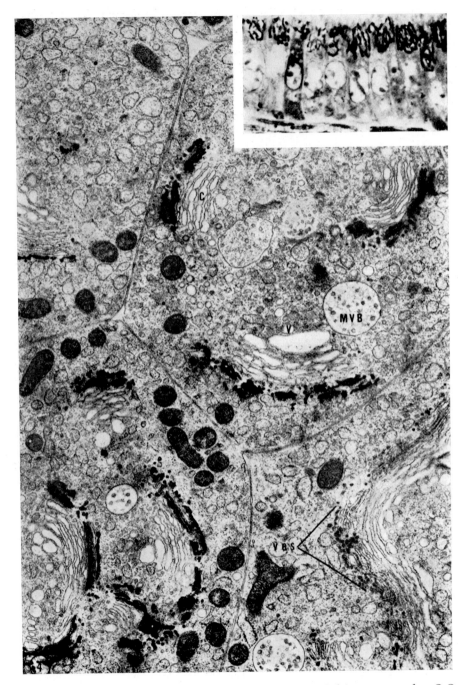

Fig. 28. Portions of cells from the mouse epididymis showing Golgi apparatus after OsO$_4$ impregnation. The electron micrograph shows discontinuity in staining. The inset, which is a light micrograph, shows the classical picture. $C$ = cisternae, $V$ = vacuole, $MVB$ = multivesicular body, $VES$ = vesicles. From FRIEND and MURRAY, Amer. J. Anat. **117**, 1965. ×10,800.

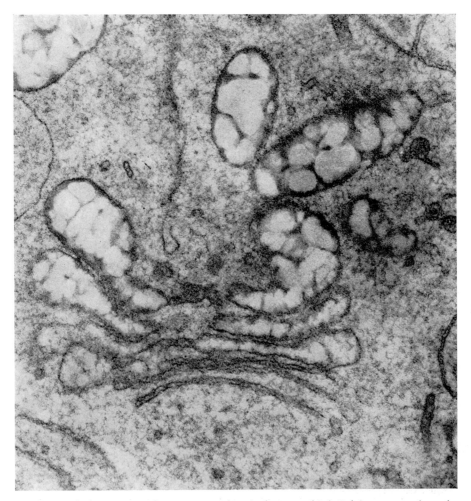

Fig. 29. Particulate masses of secretory product in hypertrophied Golgi apparatus in maize root cells. From WHALEY et al., in: Cellular membranes in development (LOCKE, M., ed.). New York: Academic Press. 1964. ×50,000.

specialization there must be local differentiations in cisternal membrane structure, because in some instances Golgi apparatus may produce more than one material simultaneously (COULOMB and COULON 1971, SAGE and JERSILD 1971, MIGNOT et al. 1972, OVTRACHT and THIÉRY 1972, DAWSON 1973).

Among the evidence for membrane changes it should be observed that the osmiophilia of the cisternae may decrease from the proximal to the distal face. This observation provides, by the way, an interesting example of the refinement of techniques since GOLGI's day. GOLGI's demonstration of the apparatus depended upon its general impregnation. Now one face of the organelle sometimes can be distinguished from the other on the basis of differences in osmiophilia. FRIEND and MURRAY (1965) have demonstrated the progress of technique strikingly by combining an optical micrograph and an electron micrograph of the same type of cell (Fig. 28). In the optical micrograph one sees an apparently continuous Golgi structure. In the electron micrograph one sees more or less discontinuous Golgi stacks with greater osmiophilia toward the proximal face.

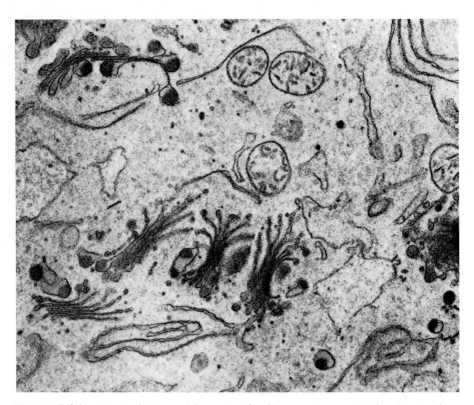

Fig. 30. Golgi apparatus in a transition zone of a *Zea mays* root apex showing one form of aggregation of secretory product within another. From H. H. MOLLENHAUER, J. Ultrastruct. Res. **12**, 1965 b. ×25,000.

In some materials specific fixatives including permanganates may demonstrate differences in cisternal contents across the apparatus and may even distinguish secretory products in particulate form within the cisternae (Fig. 29) (MOLLENHAUER and WHALEY 1963, WHALEY *et al.* 1964). Sometimes the particles show distinctive paracrystalline or crystalline structure (for example, see MANTON 1966 a, b, SILVEIRA 1967). In a few instances (Fig. 30) highly localized membrane differentiations are reflected in one type of inclusion being found within another type of vesicle.

Several of the morphological features that distinguish one face of the Golgi apparatus from the other sometimes can be made strikingly apparent in reconstruction of Golgi stacks (Fig. 31). These include the changes in cisternal dimensions, localized hypertrophy, or aggregations of secretory products.

In general, relatively few techniques have been applied to demonstrations of the various activities of the Golgi apparatus, and since BOWEN's work (see BOWEN 1929) much of the concentration has been on techniques that would demonstrate involvement in secretion. An exception is a small number of cytochemical techniques useful mainly in localizing components that go

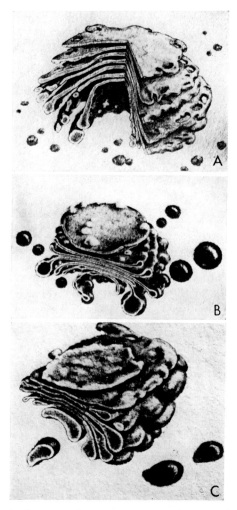

Fig. 31. Drawings made from serial section reconstructions of the Golgi apparatus of several cell types in the maize root, (A) from the meristematic cell (B) from an epidermal cell and (C) from a rootcap cell. All show changes in cisternal size, degree of fenestration, and buildup of secretory product from the proximal face (top) toward the distal face (bottom). From J. E. KEPHART, Cell Research Institute, University of Texas at Austin.

into the lysosomal system. Recently there have been some applications of other impregnation techniques, among them an osmium-zinc iodide (OZI) procedure. This procedure gives variable demonstration to certain intra-cytoplasmic membrane systems, and there are suggestions that it may be

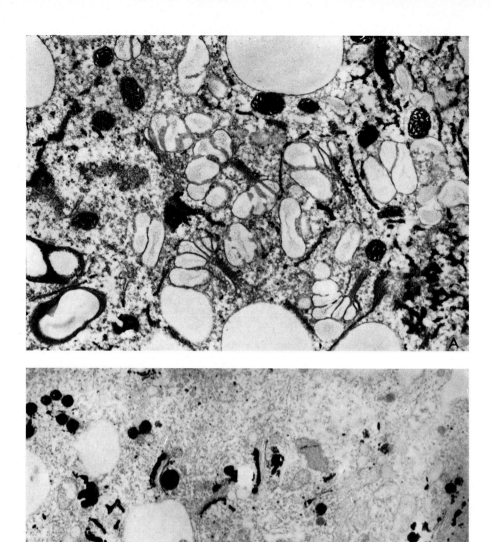

Fig. 32. Sections of *Zea mays* root apex cells showing OZI staining patterns. The conspicuous accumulations of secretory products are not stained. *A.* An outer rootcap cell. Staining of the Golgi stack is at the proximal face. Secretion vesicles are unstained. *B.* Epidermal cell. Addition of NaCl to the staining procedure has resulted in staining of both proximal and distal faces of the Golgi stacks. Secretory vesicles adjacent to the Golgi stacks show a variable amount of staining. Mature vesicles are unstained. From DAUWALDER and WHALEY, J. Ultrastruct. Res. **45**, 1973. ×15,500.

useful in distinguishing Golgi apparatus of different tissues. There are not yet enough data to assess its significance, but it differs from the more conventional procedures in at least two respects: first, it is sometimes a relatively specific stain for the more proximal cisternae, and, second, it fails to stain the accumulations of secretory products which are frequently visible in the

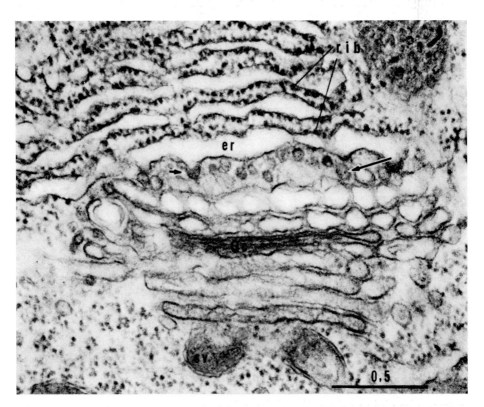

Fig. 33. An association between the Golgi apparatus and endoplasmic reticulum. Portion of a cell of a mealybug testis. *RIB* = ribosomes, *ER* = endoplasmic reticulum, *GC* = Golgi cisternae. Arrows indicate blebs from the ER towards the proximal Golgi cisternae. From WHALEY, in: Organisation der Zelle. III. Probleme der biologischen Reduplikation. (SITTE, P., ed.). Berlin-Heidelberg-New York: Springer. 1966. ×52,000

preparations (Fig. 32). Still more puzzling is that in certain tissues or under certain circumstances the OZI technique stains cisternae on both the proximal and distal faces and leaves unstained those in the midregion. This is a pattern that is not readily explained in view of its failure to demonstrate secretory products and no more readily explained as an interpretation of membrane changes across an axis (DAUWALDER and WHALEY 1973).

Ever since the Golgi apparatus was assigned a role in the intracellular events of secretion and especially since secretion came to be recognized as a general cellular process, investigators have pointed out that there must be numerous if transient associations between the apparatus and other cellular components, and there have been a few records of fairly consistent associations of this sort (see BOUCK 1965). These associations would certainly be important metabolically, and they have been discussed frequently as possibly involved in the transfer of membranes and specific cellular components (WHALEY *et al.* 1971). They have also occasionally been used as a means of distinguishing between the two faces. Fig. 33 illustrates such an association

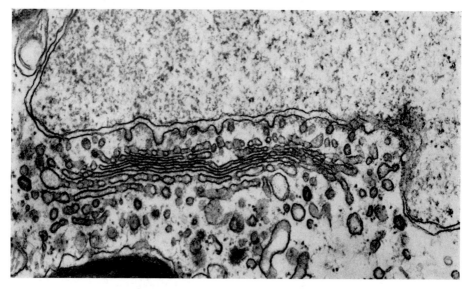

Fig. 34. Blebbing from the nuclear envelope toward the Golgi apparatus in the zoospore of an alga, *Tribonema vulgare*. From MASSALSKI and LEEDALE, Br. phycol. J. **4**, 1969.
×42,000.

between the Golgi apparatus and the endoplasmic reticulum. The profile of the endoplasmic reticulum adjacent to the Golgi apparatus is smooth and seems to be giving off small vesicles, whereas the other profiles are granular. The implication is one of a transfer of membrane and perhaps bounded material from the endoplasmic reticulum to the Golgi apparatus. The cells of certain lower organisms and of embryonic stages in some higher organisms are characterized by somewhat comparable associations of the Golgi apparatus and the nuclear envelope (Fig. 34) or of associations between extensions of the nuclear envelope and the Golgi apparatus (Fig. 35). Again the implication is of a transfer of material from one membrane system to another. There are three problems involved with all such interpretations: first, the dimensions of vesicles assumed to be in transit from one system to another rarely accord with the dimensions of the cisternae, and there must be significant changes in volume if membrane fusion processes are involved in the formation of cisternae in such a manner; second, several investigators (see Section XII) have shown distinct compositional differences among cytoplasmic membrane systems, and this fact with the failure of numerous experiments to demonstrate clearly membrane-bounded transfer from the endoplasmic reticulum to the Golgi apparatus has raised questions as to whether this transfer occurs in such simple form (if by chance it does, then one may have to assume progressive displacement of the Golgi cisternae across the stack or the movement of materials through a succession of membranes); and third, what seems to be the distal or secreting face of the Golgi apparatus appears in association with the nuclear envelope or endoplasmic

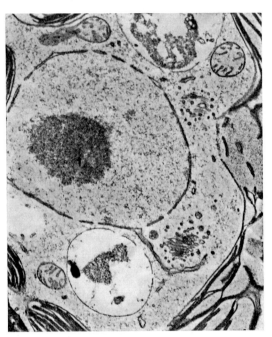

Fig. 35. Extension from nuclear envelope to the vicinity of the Golgi apparatus in an algal cell. From WALNE, Amer. J. Bot. **54**, 1967. ×16,000.

reticulum with enough consistency to raise a question of whether a proximal-face association with these membrane systems actually indicates a functional relationship.

## D. The Golgi Complex

The term *Golgi complex* has found a number of different uses. In some of the earlier literature it was used more or less interchangeably with Golgi apparatus to discuss the organelle's association with the central bodies (centrioles) or the different staining characteristics of the chromophobic and chromophilic areas of the organelle seen with light microscopic techniques. Some investigators still use it essentially interchangeably with Golgi apparatus. It will be used here to refer to instances in which localized Golgi apparatus are variously associated to form structures that appear to occupy a considerable portion of the cell and which seem to be interconnected in a number of ways.

The form of the Golgi complex often appears to relate to the form of the cell itself and to the manner in which the cell is polarized. In cells of the mouse epididymis the complex appears to be tubular (perhaps actually cup shaped but very elongated), but the individual Golgi stacks are widely spaced and not uniformly oriented. In cells of the rabbit epididymis the complex is of the same shape, but the stacks are closely enough associated to make a continuous tube with some stacks appearing within the tube itself (Fig. 36).

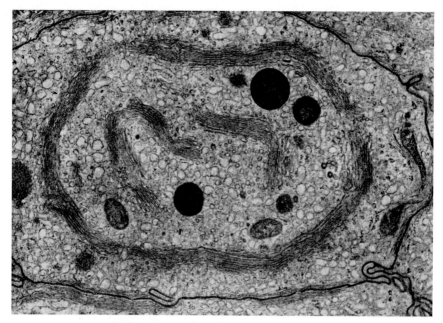

Fig. 36. Example of a Golgi complex in a cell of the rabbit epididymis. From FAWCETT, An atlas of fine structure. London-Philadelphia: Saunders. 1966. ×9,000.

In at least some cells in spermatogenesis, the complex seems to form a nearly, but not quite, continuous wall of a sphere with the proacrosomal granules which are ultimately involved in acrosome formation in a specific relationship to this structure (Fig. 37). At different stages in spermatogenesis this complex undergoes considerable modification. Simpler forms include crescents or very shallow cups (BLOCH and HEW 1960) or rings. These instances of highly organized Golgi complexes give certain characteristics to the cell with respect to the sites of accumulation and the pathways of movement of products formed in the Golgi apparatus.

## E. The Golgi Region

As applied to that portion or portions of the cytoplasm in which the Golgi stacks and associated vesicles are the conspicuous elements, the term *Golgi region* carries some of the connotations of a field in which morphogenetic or physiological changes may take place. Of concern with respect to the concept of a Golgi region is the influence of activities within the stack on the associated vesicles or the role of these vesicles in transporting substrates and possibly enzymes to the stack. It has long been supposed that secretory products undergo certain further changes after being separated from the cisternae because they often remain in close association with the stack. Whether such transformations are actually influenced by the Golgi stack is not clear. They could well be determined, as SJÖSTRAND and HANZON

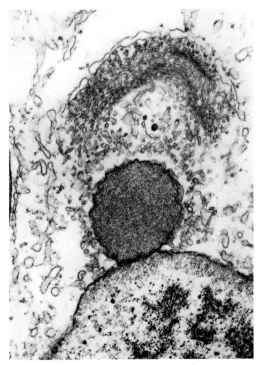

Fig. 37. Relationship of the Golgi apparatus and the acrosomal granule in a Hemipteran spermatid. From WHALEY, in: Organisation der Zelle. III. Probleme der biologischen Reduplikation (SITTE, P., ed.). Berlin-Heidelberg-New York: Springer. 1966. ×16,000.

(1954 a, b) and SJÖSTRAND (1968) postulate, by the extent to which the membranes that surround them become specialized. There are, however, some suggestions of possible influences of the Golgi region on other cellular components. One such question is inherent in the observation by HOLTZMAN et al. (1967) that what appears to be smooth endoplasmic reticulum may acquire acid phosphatase activity when it is in the vicinity of the distal face of the Golgi apparatus. Segments of endoplasmic reticulum showing such activity, presumably as a result of influences imparted from the Golgi apparatus, have been designated by NOVIKOFF and his associates as GERL. The term GERL was first used to describe an association among the Golgi apparatus, endoplasmic reticulum, and lysosomal activity (Fig. 38).

The fact that profiles of endoplasmic reticulum in association with the proximal face of the Golgi stack frequently show one membrane without ribosomes but with numerous evaginations tends to suggest that influences may extend from the other face of the Golgi region to another organelle (Fig. 33). The difficulty, of course, is that there is no certainty whether such instances as this are examples of influence being exerted by the Golgi region on other cellular components or of the other cellular components exerting influence on the Golgi region. The cytoplasm immediately surrounding any

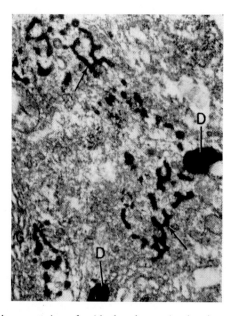

Fig. 38. Cytochemical demonstration of acid phosphatase in the elements of the endoplasmic reticulum adjacent to the Golgi apparatus in a rat ganglion, GERL (arrow). $D$ = dense bodies. From HOLTZMAN *et al.*, J. Cell Biol. **33**, 1967. Courtesy of Rockefeller University Press. $\times 27,000$.

organelle or inclusion may have certain characteristics related to the activities of that organelle or inclusion that are not shared to the same extent by the remainder of the cytoplasmic matrix. This must be true even though there are no morphological indications of it. In the case of the Golgi apparatus the undetected differentiation of the surrounding cytoplasm may be different on the opposite faces and it may vary in extent and degree with the activity of the apparatus. Thus, there is a good reason for giving consideration to a Golgi region as well as to the detectable membrane-bounded units. The limits of such influences which would define the boundaries of the Golgi region cannot at this stage be set forth. The possibility that a certain region or regions of the cytoplasm should be designated as Golgi regions also has implications discussed in later sections on the origin and replication of the organelle.

# VI. Functions

## A. The General Pattern of Functioning

Electron microscopy and allied techniques have greatly extended our knowledge of the form of the Golgi apparatus. One must also face the question of the extent to which new techniques have confirmed and expanded our knowledge of its functioning.

The early studies of the Golgi apparatus which led ultimately to recognition of its functioning in the intracellular events that are part of secretion carried with them an idea that large masses of material were evolved from the organelle. With the advent of electron microscopy it became clear that these masses of material were all membrane-bounded. By implication it is assumed that in some manner the organelle must be able to assemble and extend membranes to encompass the secretory materials accumulated within and released from the apparatus. Some reviewers still refer to its principal role as that of packaging secretory products (BEAMS and KESSEL 1968). It is becoming increasingly clear, however, that a whole range of cellular activities and probably cellular associations are dependent upon genetically controlled dynamics of cellular membrane systems in which the Golgi apparatus plays a central if little studied part.

Thus in considering the functioning of the Golgi apparatus one has to be concerned with the assembly, extension, and specialization of membranes, with its classic role in secretion, and, as will become apparent, with some other roles as well. The interrelationships among these functions are sometimes very obscure but there are enough common features so that one can select certain aspects of activity and avoid compiling a repetitious catalogue of its role in different sorts of cells.

Secretory products are, in general, combinations of proteins and lipids, proteins and carbohydrates, or all three. The membranes are likewise constructed of molecules from the several classes of components. The membranes are subject both to progressive specialization and molecular exchanges which may control their capacity to develop to the essential stage of secretory product accumulation and/or synthesis. Little is actually known about the interrelationships between product development and membrane specialization. The envelopment of secretory products by membranes appears to be invariable whether the products are assembled within the cisternae of the Golgi stack or in vesicles already separated from the stack. One explanation of this compartmentalization of product within a membrane is that it prevents potentially active substances from affecting cytoplasmic metabolism. Another is that the membranes provide for fusion with the plasma membrane and thus facilitate the passage through it of substantial amounts of material. It is possible, of course, that the enzymes which catalyze the development of particular products are intimately associated with the structural aspects of the membrane and that the association is thus a developmentally critical one.

In a gross way, one can separate materials destined to become portions of secretory products from those which become components of membranes but the turnover of all the classes of molecules included is great enough so that one has to assume at least some continuing exchanges with metabolic pools (SIEKEVITZ 1972). The ultimate site of a large portion of the Golgi apparatus product, membranes and secretory materials alike, is the plasma membrane and, transiently at least, the surface of the cell. The general evidence indicates substantial modifications of such external material both from influences within the cell and various external influences. It would seem, nonetheless, that a substantial number of the characteristics of the cell surface are genetically determined and mediated through the Golgi apparatus in the attributes of transferred membrane or secretory products or both.

Secretory products (which must be looked upon as potentially active substances plus associated materials) are formed in a sort of vital assembly line process which includes the activities of various components of the cell. Steps in this process take place in a specific sequence. In general, they represent stages in metabolism which have been compartmentalized to assure a high degree of efficiency, and they provide an example of the progressive structuring of molecular components characterized by functional roles and also by varying degrees of importance as carriers of biological information. Steps in the sequence take place at sites progressively removed from the direct action of nucleic acids. And yet in many cases they appear to be rigidly genetically controlled. The exceptions, or the experimental modifications that can be made to cause exceptions, may ultimately be keys to malfunctioning in development or to coordination in the organism. The genetic control of the formation of proteins has probably been the most notable biological advance of the 20th century and inasmuch as it is now a feature of all textbooks it hardly needs review and documentation here. Proteins are, however, components of both the membranes of and the secretory products assembled in the Golgi apparatus. A few remarks about the most fundamental stages in this development seem to be in order.

In most cases the genetic information of the cell is carried by DNA—concern here will not be either with RNA viruses or considerations of reverse transcription though the latter may be of importance in normal as well as abnormal cellular phenomena (see Section XII). This information is passed to RNA and moved out of the nucleus into the cytoplasm. RNA exists outside the nucleus in several different forms. A specific interaction between messenger RNA and ribosomal RNA (and also involving transfer RNA), in some association with ribosomal proteins, results, depending on how the RNA has been coded, in the formation of a polypeptide chain—a strand of amino acids in a genetically determined sequence. Many, but not all, polypeptide chains are formed by so-called polysomes—ribosomes moving in sequence along a strand of messenger RNA. Most polysomes are attached to the rough endoplasmic reticulum, and by some mechanism the polypeptide chains thus formed are passed through the membrane of the endoplasmic reticulum into the lumen. Here, the polypeptides are formed into proteins, apparently different ones at different loci, suggesting spatial differentiation of the endoplasmic reticulum (see also ERIKSSON et al. 1972). The processes involved may include linking of separate chains or conformational changes. Addition of components also occurs, particularly if the protein is destined to have carbohydrate side chains (glycoproteins or other protein-polysaccharides). For one class of glycoproteins the sugar residue which forms the link to the polypeptide chain is N-acetylglucosamine, and it is characteristically bound to asparagine groups (other linking sugars may be bound to serine, hydroxylysine or other amino acids). Several studies have indicated that some N-acetylglucosamine may be added at the time the polypeptide chains emerge from the ribosomes. Thus, the material entering the endoplasmic reticulum bears some carbohydrate moieties that are a result of very early additions in the assembly process.

The amount of potential variability at this stage is great, controlled by genetic differences at the polysomes and in the endoplasmic reticulum. One thus has a basis for the formation of a large number of different sorts of products. One may assume a potential for increase in the diversity of products related to increases in the complexity of the genome.

In certain glycoproteins mannose may occur close to the linking sugar. In a radioautographic study of the thyroid cell WHUR et al. (1969) showed that the label from $^3$H-mannose could first be localized in the endoplasmic reticulum and subsequently traced to other cellular sites. The implication is that sugars close to the polypeptide chain may be incorporated into the assembly product in the endoplasmic reticulum and thus fairly early along the biosynthetic pathway. By contrast, some other sugars, more N-acetyl-glucosamine, galactose, fucose, and sialic acid appear to move fairly directly into the Golgi apparatus. In instances where assembled products have been analyzed, these latter moieties tend to appear as terminal groups and the evidence suggests that synthetic processes involving their incorporation are localized in the Golgi apparatus. As with the other carbohydrate moieties their points of attachment appear to be consistent and genetically determined (for details and references see SPIRO 1970, ROSEMAN 1970, HEATH 1971, SCHACHTER and RODÉN 1973).

There is a good deal of evidence from literature not related to the Golgi apparatus (see in AMINOFF 1970) that terminal carbohydrate groups are capable of adding definitive specificities. If this evidence and material discussed in later sections are valid then a genetically controlled assembly of information-carrying macromolecules is completed in the Golgi apparatus where certain specificity characteristics are added. Thus the formation of much-branched, highly specific compounds appears to be under the control of the genome and its modifiers from its earliest stages to completion. The lipid groups that become part of conjugated compounds are apparently synthesized in the endoplasmic reticulum, i.e., fairly early in the assembly process.

This pattern suggests that complex materials that become important extracellularly are formed intracellularly under the influence of the genes. The various groups associated with these macromolecules may serve protective purposes or may add various characteristics to membrane surfaces. DISCHE (1966) has repeatedly pointed out that conjugated compounds may have enhanced capacity for carrying biological information and that their frequent presence on cell surfaces may increase diversity and impart certain specificities.

In assembly, extension, and specialization of membranes and in the various synthetic processes involved in the formation of products within the compartments they constitute, there necessarily has to be a varying collection of enzymes and coenzymes in particular configurations if all the genetic control at these stages has been passed from nucleic acids to enzymes. There are also suggestions of the presence in the Golgi apparatus of some enzymes which seem to be effective in catalyzing reactions that take place outside the Golgi apparatus *per se,* and there is extensive proof of the presence of enzymes which contribute to the lysosomal system and not to secretion in the usual sense, though some of them may ultimately be active outside the cell.

It is apparent that the organelle functions in a complex manner in maintaining a dynamic balance between various phases of metabolism. The succeeding sections are intended to emphasize the basic pattern and some of the better known variations. They provide proof that the Golgi apparatus is a major functional organelle and as such dependent upon and capable of affecting the other components of the cell.

## B. The Pancreatic Exocrine Cell

Because of its polarity, its cyclic secretion, and other factors, the pancreatic exocrine cell was an early choice of investigators concerned with activities of the Golgi apparatus (CAJAL 1914). Zymogen, which is a general term applied to the secretory product of this and some other glands, in this instance contains proteases, lipases, carbohydrases, and nucleases. These enzymes are inactive while membrane-bounded in the cytoplasm but become active following exocytosis. An exception is the pathological condition called pancreatitis. In this condition the enzymes may become active while within the cell itself and actually bring about its breakdown. Enzyme secretion is ordinarily induced by feeding, but may be induced experimentally. Cyclically the cell is emptied of zymogen granules and then builds up a new accumulation for secretion at the next feeding.

Maturation of the zymogen in the so-called condensing vacuoles can be followed in terms of both size and the density with which the granules stain (Fig. 39). Most investigators have attributed this maturation to continued proximity of the condensing vacuoles to the Golgi apparatus. SJÖSTRAND and HANZON (1954 a, b, see also SJÖSTRAND 1968), who studied the buildup of zymogen in unborn mice where it is not subject to secretory cycles, have attributed maturation instead to continued changes in the membranes after they are separated from the Golgi cisternae. They hold that in the normal exocrine cell condensing vacuoles are kept close to the Golgi apparatus only by the continued accumulation of zymogen granules between secretory cycles and that maturation of zymogen at this location is fortuitous. Exocytosis is brought about by two specific hormones: secretin and pancreozymin (BLOOM and FAWCETT 1968). Stimulation of the vagus nerve will also bring about the secretory activity.

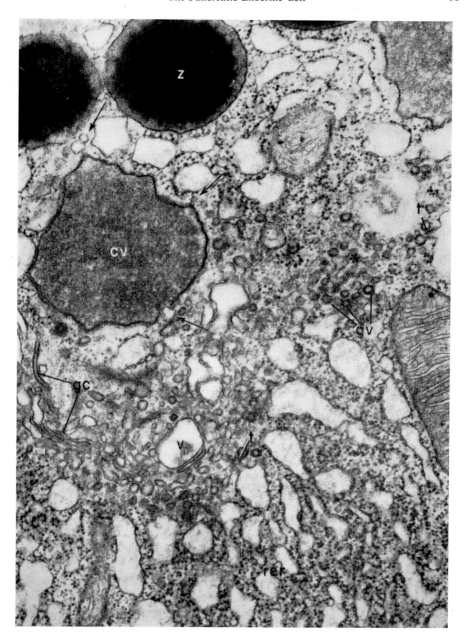

Fig. 39. Section of a pancreatic exocrine cell from a guinea pig showing the Golgi complex. The rough endoplasmic reticulum (*rer*) shows smooth-surfaced buds (*t*) projecting toward the Golgi complex. Smooth-surfaced vesicles (*gv*) occasionally fuse to form alveolate tubelike structures (\*). Small Golgi vacuoles (*v*), Golgi cisternae (*gc*), large condensing vacuole (*cv*) with smooth-surfaced blebs (arrows), and zymogen granule (*z*). From JAMIESON and PALADE, J. Cell Biol. **34**, 1967 a. Courtesy of Rockefeller University Press. ×39,000.

The detailed electron microscopy story of exocrine cell secretion was first worked out in a series of experiments by members of PALADE's group and it became a model for secretory-product formation (see PALADE 1966) until it was shown that there are substantial differences in some instances of secretion. Association of the manufacture of secretory proteins in the pancreatic exocrine cell with the ribosomes was lent validity by a series of biochemical studies by SIEKEVITZ and PALADE (for a summary see PALADE 1966) which ultimately concentrated on one of the zymogen granule components—α chymotrypsinogen. They found that shortly after the injection of $^{14}$C-leucine, radioactivity was concentrated in ribosomes attached to microsomes derived from rough endoplasmic reticulum. Analyses made at specific time intervals after injections suggested that radioactive material moved into the lumen of the endoplasmic reticulum and ultimately appeared as a component of the chymotrypsinogen in the zymogen granules. This suggestion of intracellular movement of a secretory protein led to a further series of experiments, the principal one of which was a radioautographic study carried out by CARO and PALADE (1964) and involved the introduction of $^{3}$H-leucine and a study of its distribution at specific time intervals after incorporation. At short intervals after labelling, 3 to 5 minutes, most of the label appeared over the rough endoplasmic reticulum in the basal region of the cell. At 20–40 minutes, most of it appeared in association with the Golgi region—notably the so-called condensing vacuoles characteristically associated with the distal face. After an hour, most of the label was seen over zymogen granules. These observations coupled with the long-known cyclic secretion of the zymogen granules and related to other information concerning the basic steps in the synthesis of proteins led to a definitive conclusion that the synthesis takes place at the ribosomes and that the endoplasmic reticulum is involved. Secretory proteins are then transferred from this membranous system to the Golgi apparatus, undergo a series of changes in the condensing vacuoles, and are finally discharged into the pancreatic duct (Fig. 40). [In most of the work on the pancreas the carbohydrate moiety of zymogen, which amounts to about 2%, was largely neglected. The synthesis of mucopolysaccharides which might aid somehow in the packaging of the digestive enzymes could indicate a further involvement of the Golgi apparatus (BERG and YOUNG 1971).]

Whereas BOWEN (1929) was correct in ruling out the nucleus and the ergastoplasm from participation in direct conversion into components of secretory products, these organelles play a part in the basic determination of the character of these products and are responsible for the formation of molecular skeletons which can be further elaborated to form the complex end products. This further elaboration does in many cases, as BOWEN supposed, take place in the Golgi apparatus or in vesicles derived from it.

Conspicuously missing is information showing how the proteins are transferred from the lumen of the rough endoplasmic reticulum to the smooth membrane-bounded cisternae of the Golgi apparatus or to membranous units presumably derived therefrom. The association of many small vesicles with the cisternae of the Golgi apparatus and/or in the vicinity of the condensing vacuole makes it appear from the micrographs that there might be a transfer via membrane-bounded vesicles. JAMIESON and PALADE (1966, 1967 a, b) investigated this question in some detail, postulating that such vesicles might arise from the endoplasmic reticulum, the Golgi apparatus, or might merely act as *shuttle vesicles*. JAMIESON and PALADE (1968 a, b, 1971 a) were further able to uncouple the synthesis of protein from its transport from the rough endoplasmic reticulum to the Golgi apparatus. They found that transport would continue when they blocked protein synthesis with cyclohexamide. They found the transport insensitive to glycolytic

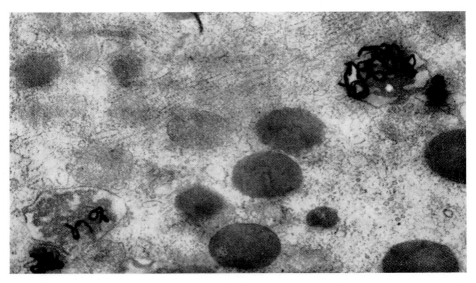

Fig. 40. Portion of a pancreatic exocrine cell from a rat showing the concentration of label from ³H-leucine over the condensing vacuoles twenty minutes after injection. From CARO and PALADE, J. Cell Biol. **20**, 1964. Courtesy of Rockefeller University Press. ×26,000.

inhibitors such as fluoride and iodoacetate but blocked by respiratory inhibitors such as cyanide and antimycin A. The blocking by cyanide could be reversed. Their full data indicated that the transport from the endoplasmic reticulum to the Golgi apparatus and from the Golgi apparatus to the cell surface is semiseparate from the assembly of the secretory product and that at least some of the transport is supported by the oxidation of long-chain fatty acids. They did establish a case for quantal transfer rather than for transfer of individual molecular units. Later, MELDOLESI and coworkers (MELDOLESI *et al.* 1971 a, b, c, MELDOLESI and COVA 1971, 1972 a, b) were to complicate this matter further by demonstrating distinct compositional differences between the membranes of the endoplasmic reticulum and those of the Golgi apparatus and thus questioning the possibility or extent of intracellular membrane mixing.

There still is not a firm explanation of how material is transferred from one system to the other but it may be important to note that material does not usually accumulate in the Golgi apparatus or its associated vesicles until

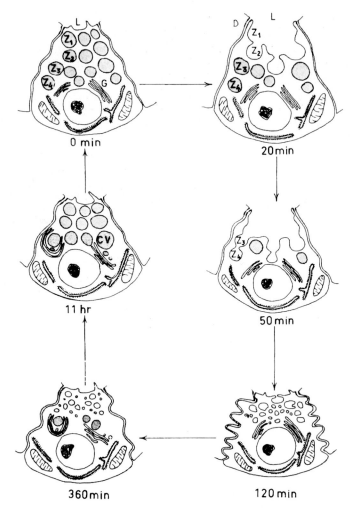

Fig. 41. Schematic diagram of a rat parotid gland cell in cyclic zymogen secretion at various intervals after feeding. $Z_1$–$Z_4$ represent zymogen granules fusing in sequential order. $G$ = Golgi apparatus, $L$ = lumen, $CV$ = condensing vacuole. From AMSTERDAM et al., J. Cell Biol. **41**, 1969. Courtesy of Rockefeller University Press.

after the membranes of these structures have reached a certain level of development.

That the assembly of secretory products and their transport are relatively independent was recognized long ago. It was BOWEN's (1929) reason for trying to define one set of intracellular processes as secretion and the discharge of the products as excretion. It does seem clear that they should be regarded as separate phases of cellular activity.

Zymogen is also produced by the parotid gland and some useful studies of its formation and transport generally confirming PALADE's interpretation have been made by AMSTERDAM et al. (1969) (Fig. 41). ICHIKAWA (1965)

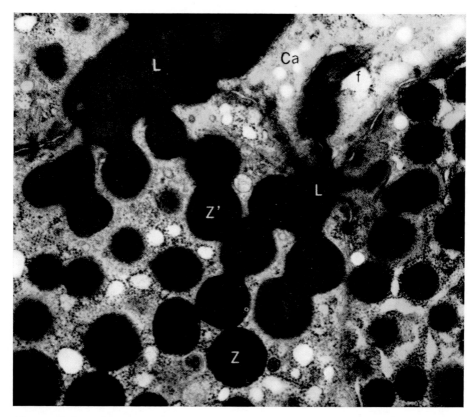

Fig. 42. Fusion of zymogen granules after contact with the luminal membrane of a canine pancreas cell resulting in mass exocytosis. From ICHIKAWA, J. Cell Biol. **24**, 1965. Courtesy of Rockefeller University Press. ×23,000.

has made an interesting series of observations on the movement of zymogen granules in the canine pancreatic exocrine cell. His observations suggest that while free in the cytoplasm such granules do not fuse but if one granule comes in contact with the plasma membrane there is a fusion which may also involve any granules in contact with this one, resulting in a mass exocytosis and deep invaginations of the plasma membrane into the cell (Fig. 42).

Immediately after the feeding process the extent of plasma membrane is seen to be increased greatly. There follows some sort of compensation by which the extent of plasma membrane is reduced to its prefeeding stage. This compensation has been variously envisioned involving movement of small membrane-bounded vesicles back to the endoplasmic reticulum and/or the Golgi apparatus (PALADE 1959), movement of nonvisible molecular components (FAWCETT 1962), and movement of various membrane components in different stages of organization (HOKIN 1968) (Fig. 43).

The question arises whether these membranes or membrane components are reutilized in a *recycling* process to form new membranes. Evidence to the contrary has been presented by AMSTERDAM *et al.* (1971) showing that in the parotid gland new membrane protein is formed concurrently with new secretory protein. In addition, the evidence already cited from various investigators that cellular membranes differ in composition (see MELDOLESI *et al.* 1971 a, b, c) makes direct recycling of membrane segments unlikely. There might, however, be some sort of selective fusion/fission process involving limited membrane regions which would not be reflected in gross compositional differences.

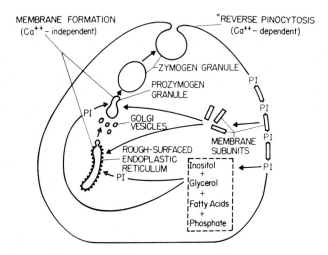

Fig. 43. A possible concept of membrane circulation involving different building block forms of membrane sub-units. *PI* = phosphatidylinositol. From HOKIN, Int. Rev. Cytol. **23**, 1968.

Evidence in favor of recycling has been presented by JAMIESON and PALADE (1971 a, b) in studies of the transport of pancreatic zymogen. Their data suggest an extensive reutilization of intracellular membranes or membrane components. There are many data on membrane movement toward the plasma membrane but little is actually known about the pathway of returning membrane. It may be that some membrane components (perhaps modified in the lysosomes, see HOLTZMAN *et al.* 1971) are subject to various utilizations in the metabolic pools of the cytoplasm and not immediately reutilized in membrane building.

Whereas compensatory mechanisms for membrane removal and membrane recycling are questions to be considered for most cells with a high level of secretory functioning, in certain rapidly growing cells membranes transferred from the Golgi apparatus to the plasma membrane appear to remain in that location. Two examples are the rapidly extending surface membranes of the growing pollen tube in which the Golgi apparatus can be seen to be very active (LARSON 1965) (Fig. 44) and what appears to be the formation of new membrane in the initial stages of cell plate formation (see Section VII, B).

In the guinea pig and the dog, under certain experimental conditions (PALADE 1956, ICHIKAWA 1965, FAWCETT 1966), there sometimes appear in the lumen of the rough endoplasmic reticulum, granules of the same size and density as the zymogen granules. This has led FAWCETT (1966) to suggest that the endoplasmic reticulum membranes may sometimes share metabolic capacities with those of the Golgi apparatus.

The importance of the pancreatic exocrine cell lies in certain of the details that it provides as a model of secretion. Study of this cell type has revealed the sequence of sites at which secretory proteins are assembled and indicates that they undergo a progressive maturation during this sequence. These

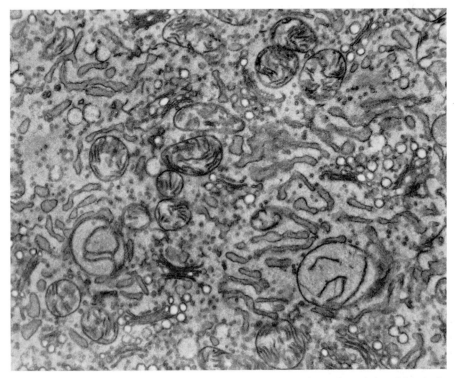

Fig. 44. Golgi apparatus very active in the production of membrane-bounded vesicles during pollen tube growth in *Parkinsonia aculeata*. From LARSON, Amer. J. Bot. **52**, 1965. ×18,000.

studies have demonstrated that a cell can form large amounts of secretory product for export in a relatively short time, have shown the importance of the surrounding of active products by a membrane and the participation of that membrane in exocytosis of material, and have illustrated the dynamic character of the surface membrane in a system in which secretion is cyclic. Further investigations have suggested that different aspects of cellular activity are involved in the movement of the secretory material into the Golgi apparatus and from the Golgi apparatus out of the cell. Though other aspects of the study of the pancreatic exocrine cell have been important, its greatest significance will probably be as an example of the manner in which different techniques utilized by different investigators first demonstrated the formation and assembly of protein making up the bulk of a secretory product. It is of interest that the successive steps involved include activity on the part of each of those components considered by the light microscopists to be of concern but not made identifiable until the advent of electron microscopy and its allied techniques.

## C. The Goblet Cell

Epithelial cells of the digestive tract were important in the early investigations of the Golgi apparatus because of their distinctive polarity and the intensity of their secretions. These secretions include both digestive enzymes and mucosubstances; the latter are important in lubrication, in the adherence of food particles to one another, and probably in the intimacy of association between the enzymes and the food particles. Ito and Winchester (1963) have prepared a series of diagrams of different gastric glands in the bat (Fig. 45) which illustrate much about differences in secretion. For a comprehensive treatment of the ultrastructure of cells from various regions of the digestive tract see Toner et al. (1971).

Concern here will be with the goblet cell as it is a particularly good example of a quite different set of intracellular processes involved in secretion than are illustrated by the pancreatic exocrine cell. The goblet cell of the colon was recognized as producing mucigen and doing so in a very specific pattern as long ago as Cajal's famous experiment in 1914. Peterson and Leblond (1964 a, b) selected colonic goblet cells for investigation by electron microscopy and radioautography because these cells secrete large amounts of material with a high proportion of carbohydrate. Eylar (1965) made a suggestion that it was the presence of carbohydrate that triggered the exocytosis of secretory materials. This is probably not correct but presence of some carbohydrates seems to explain the involvement of the Golgi apparatus. A colonic goblet cell is illustrated in Fig. 46. It has a normal active life of 3 or 4 days after which it is replaced by newly maturing cells. It is highly polarized, with the Golgi apparatus forming the base of the goblet just above the nucleus and the goblet characteristically being filled with enlarging and otherwise changing mucigen granules. These granules are progressively displaced toward the lumen of the intestine into which they are discharged in a form of secretion. Toner et al. (1971) emphasize that there always is a confluence of mucus after discharge from the cell.

Both in their earlier papers Peterson and Leblond (1964 a, b) and in later papers Neutra and Leblond (1966 a, b, 1969) give a detailed description of the Golgi apparatus in these cells. They point out that the more proximal cisternae are characteristically immediately adjacent to segments of rough endoplasmic reticulum and rarely show hypertrophy or swelling in any region. This latter development occurs quite suddenly at some distance across the stack. When it does occur, it leads to the conversion of the more distal cisternae completely into mucigen granules. In time-sequence studies they have noted that secretion in this instance appears not to be cyclic but continuous and that a distal cisterna is converted into mucigen granules approximately every 2 to 4 minutes. This conversion is compensated for by

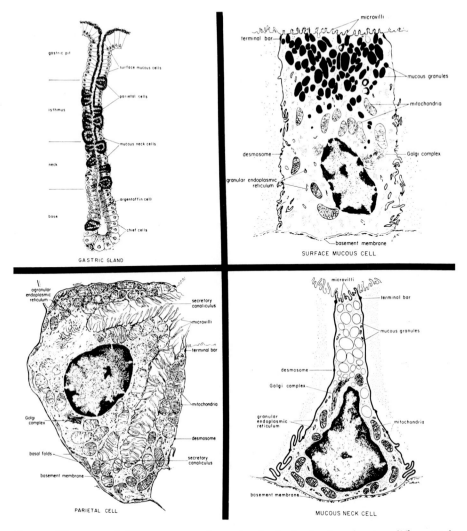

Fig. 45. Diagrams of different types of gastric glands in the bat showing differences in mucosal secretion. From Ito and Winchester, J. Cell Biol. **16**, 1963. Courtesy of Rockefeller University Press.

the formation of a new proximal cisterna at the same rate so that the number of cisternae per stack does not change with the progress of secretion. There is, however, evidence (Toner *et al.* 1971) that Leblond's interpretation may not be entirely correct. The filling of the goblet with mucigen granules implies that secretion exceeds the limit necessary to maintain a balanced state. Hence at some phase the activities are less balanced than represented by Leblond for the mature cell.

Such an interpretation implies movement of the cisternae across the stack with #1 becoming #2, #2 becoming #3, etc. and the distalmost cisterna(e) being completely converted to mucigen granules. This behavior of the organelle has also been suggested by MOLLENHAUER and MORRÉ (1966) and BROWN (1969) in quite different types of cells. The mucigen granules increase in size and are progressively displaced until ultimately they reach the apical membrane. LEBLOND has interpreted this progressive displacement of cisternae as a typical Golgi apparatus pattern though his position is that the rate of displacement may vary greatly. This pattern will explain certain aspects of the behavior of the Golgi stack, but it is difficult to understand how dis-

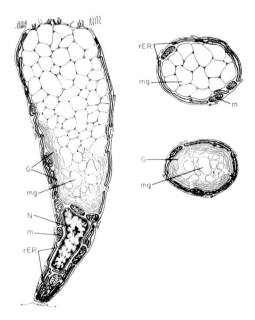

Fig. 46. A diagrammatic interpretation of the structure of the goblet cell as seen in longitudinal view, at the left, and at two different levels; through the Golgi apparatus, right lower, and over the Golgi apparatus, right upper. *rER* = rough endoplasmic reticulum, *m* = mitochondrion, *N* = nucleus, *mg* = mucigen granule, *G* = Golgi apparatus. From NEUTRA and LEBLOND, J. Cell Biol. **30**, 1966 a. Courtesy of Rockefeller University Press.

placement of cisternae could occur in those Golgi apparatus with inter-cisternal structures or either how or why it should take place in instances in which the discharge of secretory material is by quite different processes.

In their earliest experiments, PETERSON and LEBLOND (1964 a, b) used ³H-glucose, some labelled at the 1 position and some at the 6 position, for incorporation into the goblet cells. They found an early distribution of glucose label over the cisternae of the Golgi apparatus and interpreted their data to indicate that it moved progressively from the more proximal to the more distal cisternae and then to the lumen of the intestine. Although polysaccharide synthesis had been suggested as being associated with the Golgi apparatus from the earlier period of studies, this was direct evidence from radioautographic labelling.

There seemed to be no prior distribution of the labelled glucose to other cellular sites though this could not be absolutely ruled out with the time

Table 1. *Incorporation of Carbohydrate Precursors in the Golgi Region of the Rat*

| Tissue and cell | Radioautographic reaction in Golgi region 5 to 15 min after injection of | | |
| --- | --- | --- | --- |
| | Glucose-H$^3$ | Galactose-H$^3$ | Sulfate-S$^{35}$ |
| Small intestine | | | |
| goblet | +++ | — | +++ |
| columnar | ± | ++ | — |
| Large intestine | | | |
| goblet | ++++ | ++ | +++ |
| columnar | ± | + | — |
| Brunner's gland | | | |
| mucous | ++++ | +++ | — |
| Stomach | | | |
| surface mucous | ++++ | ± | ++ |
| mucous neck | + | — | — |
| Salivary glands | | | |
| sublingual (mucous) | +++ | ++ | — |
| submaxillary (mucous) | + | — | — |
| Trachea | | | |
| epithelial mucous | ++ | + | + |
| mucous gland | +++ | — | +++ |
| serous gland | ± | — | — |
| Pancreas | | | |
| acinar | ± | + | — |
| Liver | | | |
| parenchymal | — | + | — |
| Epididymis | | | |
| epithelial | ± | ++ | — |
| Kidney | | | |
| proximal tubule | — | ++ | — |
| Chondrocytes | | | |
| in trachea | + | + | ++ |
| in knee joint | +++ | +++ | +++ |

± in the "glucose" column indicates a reaction seen only after high dose in 100 g rat.

($^3$H-glucose and $^3$H-galactose along with the indication of sulfation in various cells of the rat. From Neutra and Leblond, J. Cell Biol. **30**, 1966 a. Courtesy of Rockefeller University Press.)

periods used. The experiment was subject to some criticism because glucose becomes a component of a number of cellular constituents. Accordingly, a more extensive experiment using both $^3$H-glucose and $^3$H-galactose was performed by Neutra and Leblond (1966 a, b). They found that both carbohydrates were rapidly concentrated in the Golgi apparatus and thereafter the labels were found associated with products of the apparatus though some cells had a preference for one sugar or the other (Table 1).

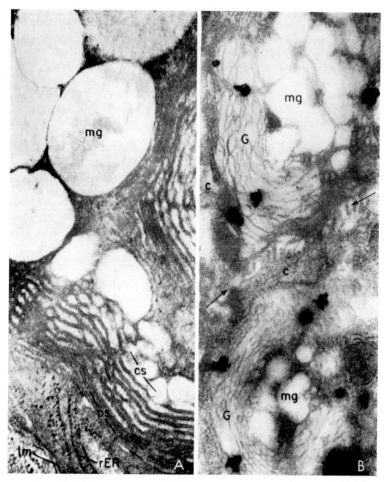

Fig. 47. Sections of colonic goblet cells of rat. $A =$ A control section showing the cell
membrane (*lm*), rough endoplasmic reticulum (*rER*), peripheral cisternae of Golgi apparatus
(*ps*), central cisternae of Golgi apparatus (*cs*), and mucigen granules (*mg*). $B =$ Radioautograph of transverse sections 5 minutes after the injection of ³H-glucose showing label
over the Golgi apparatus (*G*) and the absence of label over the cytoplasm (*c*) and the
mucigen granules (*mg*). Arrow indicates intercellular space. From NEUTRA and LEBLOND,
J. Cell Biol. **30**, 1966 a. Courtesy of Rockefeller University Press. $A \times 45,000$, $B \times 15,600$.

As an example of their findings from radioautography applied to carbohydrates rather than proteins, they found the sequence illustrated in Figs. 47
through 49. This offered a proof now repeated in many different tissues
that certain simple carbohydrates are rapidly assimilated into the Golgi
apparatus. This assimilation may be spectacular in such instances as plant
cells where the secretory products are proportionately highly carbohydrate
(see Section IV). In their 1966 experiments NEUTRA and LEBLOND were able
to confirm a general observation that sulfation also takes place in the Golgi
apparatus (see Section VI, F).

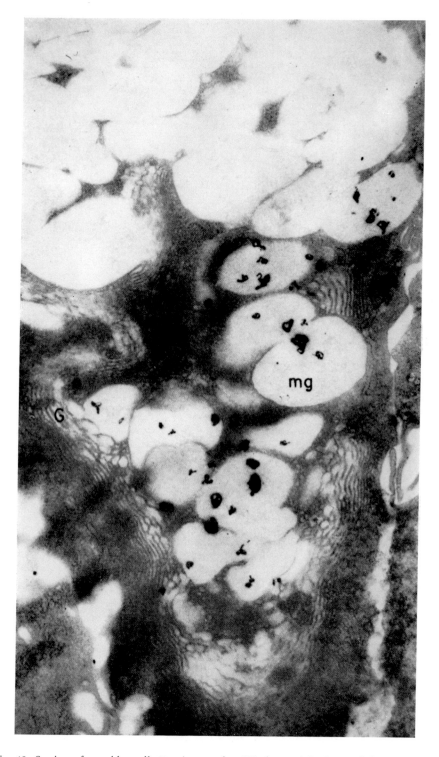

Fig. 48. Section of a goblet cell 40 minutes after ³H-glucose injection. Label is now over the mucigen granules (*mg*) adjacent to the Golgi cisternae (*G*). From Neutra and Leblond, J. Cell Biol. **30**, 1966 a. Courtesy of Rockefeller University Press. ×24,000.

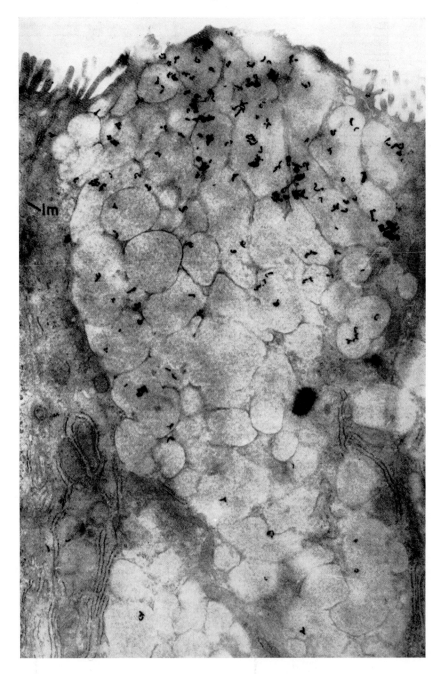

Fig. 49. Section of a goblet cell 4 hours after injection of ³H-glucose. Label is now over the more apical mucigen granules. (*lm*, lateral cell membrane.) From NEUTRA and LEBLOND, J. Cell Biol. **30**, 1966 a. Courtesy of Rockefeller University Press. ×14,500.

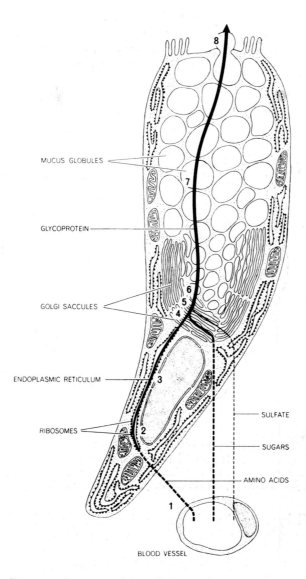

Fig. 50. A diagrammatic representation of the events in the assembly of mucus in the goblet cell. (1) Precursors move into the cell from the bloodstream. (2) Polypeptides, synthesized on the ribosomes, are polymerized into proteins and move up through the rough endoplasmic reticulum. (3) These proteins are somehow transferred to the Golgi cisternae. (4) Simple sugars move directly into the Golgi apparatus to be conjugated with the proteins. (5) The resulting glycoproteins are sulfated in the Golgi apparatus. (6) The distalmost cisternae are transformed into mucigen granules. (7) These granules are progressively displaced by the new ones. (8) The mucigen granules are secreted into the intestine. From NEUTRA and LEBLOND, Sci. Amer. **220** (2), 1969. Courtesy of Scientific American.

NEUTRA and LEBLOND (1969) have presented a diagram (Fig. 50) which represents a simplified version of the assembly process responsible for the production of mucigen. Their assumption is that precursors enter the goblet cell from an associated capillary. Amino acids are synthesized into protein in the endoplasmic reticulum and transferred to the Golgi apparatus. Sugar

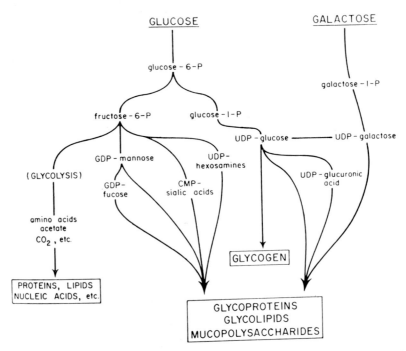

Fig. 51. A schematic representation greatly simplified showing the conversions of glucose and galactose. Whereas glucose is used in many aspects of metabolism, galactose more frequently becomes a component of complex surface carbohydrates. From NEUTRA and LEBLOND, J. Cell Biol. **30**, 1966 b. Courtesy of Rockefeller University Press.

entering the cell bypasses the other components and goes directly into the Golgi apparatus where it is complexed with protein and the glyco-protein is then sulfated to form mucigen which undergoes some further modification as it is displaced towards the apex of the cell. The goblet cell assembles a substantial amount of carbohydrate in the Golgi apparatus and sulfates some of it to form a largely polysaccharide secretory product. The pattern of exocytosis appears to differ from that of most other cells by not being cyclic but being continuous for the relatively short life-span of the cell. There may be progressive displacement of cisternae across an axis of maturation with the conversion of the distalmost cisterna(e) wholly into membrane-bounded secretory vesicles. Both glucose and galactose go into a number of cellular constituents, a fact NEUTRA and LEBLOND (1966 b) have taken into consideration in the presenta-tion of a simplified schematic representation (Fig. 51) which postulates that galactose may be largely incorporated into complex polysaccharides frequently found on the surfaces of the cell. Other carbohydrates are, however, of direct concern in the formation of secretory products and the apparent affinity of some of them for the Golgi apparatus is considered in succeeding sections.

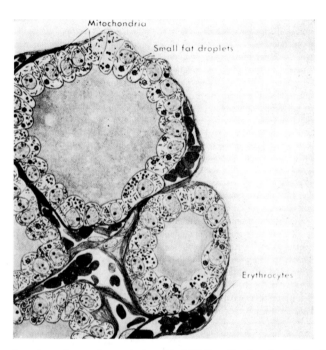

Fig. 52. A section through portions of thyroid follicle cells showing the relationship of cells to the colloid. After R. R. BENSLEY, from BLOOM and FAWCETT, A Textbook of Histology. Philadelphia-London-Toronto: Saunders. 1968.

## D. The Elaboration of Thyroid Secretions

An excellent example of the synthesis and assembly of a relatively complex protein-polysaccharide which contains numerous carbohydrate groups is seen in the elaboration of thyroglobulin, the precursor of the thyroid hormones, thyroxine and tri-iodothyronine. Thyroglobulin is incompletely assembled in the follicle cells of the thyroid gland (Fig. 52); it is then secreted into the colloid. Here its assembly is completed by iodination, and it is later taken back into the follicle cells (Fig. 53) where it is converted by lysosomal activity (see Section X, A) into the thyroid hormones. The light microscopists had correlated changes in the position of the Golgi apparatus according to whether thyroglobulin was being secreted into the colloid or hormones were being secreted into the bloodstream. The structure of thyroglobulin (noniodinated) destined for transfer into the colloid is fairly well known. The macromolecule is made up of polypeptide chains; five of them with only N-acetylglucosamine and mannose; the others have as additional

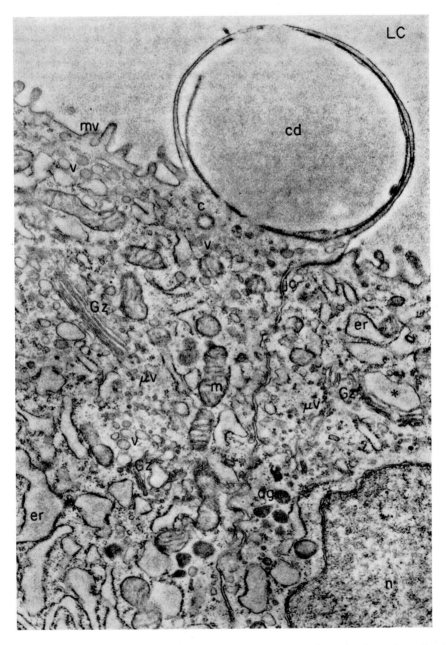

Fig. 53. Section showing the surface of a rat thyroid cell. Colloid droplet (*cd*) enclosed within an extension of cellular membrane. *dg* = dense granules, *m* = mitochondrion, *er* = endoplasmic reticulum, *c* = centriole, *v* = moderate-size vesicles, *Gz* = Golgi apparatus, *μv* = microvesicles, *LC* = luminal colloid, *mv* = microvillus, *jc* = junctional complex, *n* = nucleus, * = enlarged Golgi cisterna. This micrograph was taken from TSH-injected material. From WETZEL *et al.*, J. Cell Biol. **25**, 1965. Courtesy of Rockefeller University Press. ×21,000.

carbohydrate moieties galactose and fucose or sialic acid (see SCHACHTER and RODÉN 1973). There is evidence from the study of other materials (see other sections) that these latter carbohydrates are added to the growing macromolecule in the Golgi apparatus, and studies of thyroglobulin assembly provide clear, confirmatory evidence. The basis of thyroglobulin elaboration was shown by NADLER and LEBLOND (1955), NADLER et al. (1960) and NADLER et al. (1964) in studies following the synthesis and movement of

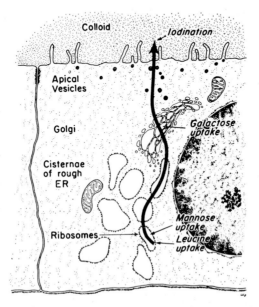

Fig. 54. Diagram of the sequential assembly of thyroglobulin showing protein and mannose association in the endoplasmic reticulum, later addition of galactose in the Golgi apparatus and iodination in the colloid. From WHUR et al., J. Cell Biol. **43**, 1969. Courtesy of Rockefeller University Press.

the proteinaceous components. Further studies indicate that the early stages of thyroglobulin assembly fit the general pattern that has already been described: synthesis of the polypeptide chain by the ribosomes and the attachment of some N-acetylglucosamine to the nascent chains as they leave the ribosomes. In a study of rat thyroid WHUR et al. (1969) showed ³H-mannose to be concentrated over the endoplasmic reticulum by 5 to 15 minutes after incorporation and at 1 to 2 hours to be transferred to the Golgi apparatus and subsequently to the colloid. This time sequence correlated closely with that for ³H-leucine incorporation; and puromycin, an inhibitor of protein synthesis, inhibited both ³H-leucine and ³H-mannose labelling. ³H-galactose was first localized (within 10 minutes) over the Golgi apparatus and subsequently over vesicles destined for secretion into the colloid. Its movement was unaffected by puromycin. The incorporation of galactose by the Golgi apparatus is commensurate with the finding of relatively large amounts of galactosyltransferase in Golgi-rich fractions of smooth membranes (BOUCHIL-

LOUX *et al.* 1970).  It is also significant that small amounts of galactose label are found over the lysosomes (see Section X, A).  A general scheme for the sequential attachment of mannose and galactose to leucine-containing chains is presented in Fig. 54.  HADDAD *et al.* (1971) subsequently followed incorporation of ³H-fucose into rat thyroids to discover that 3 to 5 minutes later it was concentrated over the Golgi apparatus.  These studies suggest that sugars close to the polypeptide chain are added early in the biosynthetic pathway, and some of these additions are carried out in the rough endoplasmic reticulum, whereas the more terminal sugars are added in the Golgi apparatus.

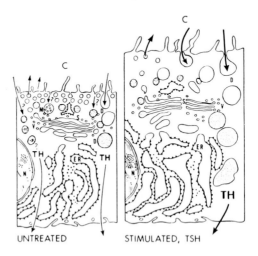

Fig. 55. Diagrams of an unstimulated thyroid follicle cell (left) and a TSH-stimulated thyroid follicle cell (right). *c* = colloid, *d* = colloid droplet, *er* = endoplasmic reticulum, *m* = multivesicular body, *n* = nucleus, *s* = Golgi saccule, *th* = thyroid hormone, *v* = Golgi apparatus-derived vesicles. From NOVIKOFF *et al.*, Fed. Proc. **23**, 1964.

Studies on the addition of N-acetylglucosamine indicate two separate sites: one in the endoplasmic reticulum and one in the Golgi apparatus (see BOUCHILLOUX *et al.* 1970, SCHACHTER and RODÉN 1973).  This is also commensurate with the known structure of thyroglobulin with one N-acetylglucosamine acting as the "linking" sugar to the polypeptide chain and one in a more terminal position.

During assembly thyroglobulin may be represented by molecules with different sedimentation rates, but by the time it reaches the colloid it is essentially a uniform material.  Its formation is completed in the colloid by the addition of iodine to tyrosine.  Iodinated thyroglobulin is stored in the colloid for varying lengths of time.  Its further movement depends upon action by the thyroid-stimulating hormone, TSH.  This stimulation produces considerable modification in the follicular cells (Fig. 55) causing endocytosis of thyroglobulin which is followed by processes involving the lysosomes and resulting in the release of thyroid hormones.

The elaboration of thyroglobulin provides an excellent example of the stepwise assembly of a particular material by functionally interrelated components of the cell. The intermediate result is a synthesis and attachment of carbohydrate side groups to a building macromolecule in a specific sequence. This process also illustrates completion of certain stages in cellular product development outside the plasma membrane of the cell. Upon proper stimulation, this membrane becomes reactive and endocytosis is an important part of the mechanism. There is then a sorting out phase under the mediation of the lysosomes and a re-secretion of products with a high degree of physiological activity, but in the opposite direction. This may account for the early observed changes in position of the Golgi apparatus in different phases of cell activity.

## E. Immunoglobulins

Immunoglobulins are apparently confined to vertebrates though other organisms sometimes appear to be characterized by surface molecules with somewhat comparable functions. The implication is that the immunoglobulins are a late evolutionary development though surface specificity relationships are at least as old as the bacteria.

Uhr (1970) has reviewed the events involved in the synthesis and secretion of immunoglobulins, primarily with immunoglobulin G (IgG) in the mouse. Referring to the work of others he points out that immunoglobulin is synthesized by plasma cells each of which produces a particular immuno-globulin in large amounts and then dies. The number of different immuno-globulins produced is a measure of the number of clones of plasma cells. Ordinarily there are many different types which together determine the individual spectrum of antibody reactivity. The following discussion of IgG is presented as an example of one type of immunoglobulin and not meant to be necessarily completely characteristic of all immunoglobulins. Immuno-globulins as a class of macromolecules differ in amino acid sequences, sugar groups, and other characteristics making for specificity as determined by the genome and reactivity to infection. IgG consists of two pairs of polypeptide chains ($H_2L_2$), one chain (H, heavy) with a molecular weight of 53,000 and a shorter one (L, light) with a weight of 22,500. IgG consistently contains about 2–3% carbohydrate. Uhr (1970) Schenkein and Uhr (1970), and Zagury et al. (1970) have elucidated the sites of addition of the carbohydrate moieties by the combination of a number of different techniques. They indicate glucosamine to be added early and to be a linking sugar. Additional glucosamine (in the N-acetyl form) and galactose are clearly added later, and both by attention to the timing and by the results of radioautographic studies they conclude that the Golgi apparatus represents the site of addition of these carbohydrate groups. Ordinarily only completed molecules of IgG are secreted though there is some evidence that individual chains may be secreted in some circumstances. According to Uhr (1970) the mechanism of assembly and secretion follows the usual pattern but the signal which brings about the secretion is not known.

Uhr concerns himself with whether the L and H chains are formed at the same or different polyribosomes. He concludes that they are formed at different polyribosomes, the heavy chain at 270 S and the light chain at 190 S polyribosomes. He quotes GooD and PAPERMASTER (1964) as reporting formation of L chains to take about 30 seconds and H chains about 75 seconds. They become linked by disulfide bonds.

Various precursor labels suggest a route consistent with that already defined from the polyribosome to the rough endoplasmic reticulum to the Golgi apparatus and then a secretion through the plasma membrane (Fig. 56).

SYNTHESIS AND SECRETION OF IG

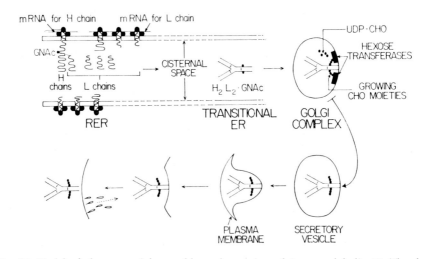

Fig. 56. Model of the sequential assembly and secretion of immunoglobulin G. The chains are joined in the rough endoplasmic reticulum where some N-acetylglucosamine is also added. There is a transfer to the Golgi apparatus and subsequently secretion in the usual manner. From UHR, Cell. Immunol. 1, 1970.

A comparison of labels associated with both free and bound polyribosomes suggests that, in agreement with other studies, most of the secretory protein is synthesized at bound ribosomes, although there is some formation at free ribosomes.

UHR (1970) considers at some length the question of IgG passage through the Golgi apparatus. He concludes that this mechanism has the advantage of a single passage through a membrane, of early separation of the product from metabolic activities in the cytoplasm, and of not adding an additional secretory mechanism to the activities of the cell. SCHENKEIN and UHR (1970) and ZAGURY et al. (1970) confirmed the results of investigations of other types of cells suggesting a sequential synthesis with the initial incorporation of some carbohydrates in the Golgi apparatus.

Basically the pattern of assembly and secretion of immunoglobulins, illustrated here by IgG, follows the pattern established for a number of secretory products. It is of particular interest because these materials constituting part of the antibody repertoire of the organism were not com-

monly thought of as secretory substances in the usual sense. The complementary information gained from the studies of thyroglobulin and immunoglobulin assembly and secretion probably provides a generalized scheme
for the synthesis of several classes of glycoproteins which are of particular
importance in the maintenance of organisms and cellular communication
within them.

## F. Sulfation

The evidence seems convincing that certain carbohydrate moieties are added
to assembling protein-polysaccharides in the Golgi apparatus. It is less clear
what function the Golgi apparatus may play in the progressive assembly of
lipoprotein or glycolipid secretory materials. As we shall see later, some
transformations seem to be involved. The Golgi apparatus, however, seems
to play still another role with the addition of sulfate to many secretory
products. Sulfation so far as it concerns the activity of the Golgi apparatus
has been most carefully worked out for the mucopolysaccharides of gastrointestinal epithelial cells and of chondrocytes.

An early observation that sulfate—$^{35}$S—is taken up into the mucus of the
goblet cells of the intestine was made by BÉLANGER (1954 a). Subsequently
JENNINGS and FLOREY (1956) localized the region of uptake to the supranuclear region which they believed to represent the Golgi apparatus. This
finding was confirmed by electron microscopy radioautography by LANE
et al. (1964). NEUTRA and LEBLOND (1966 b) examined the uptake of $^{35}$S
(Table 1) radioautographically in a time sequence in several cell types. By
their procedures, NEUTRA and LEBLOND also concluded that it was the Golgi
region of the goblet cell in which the uptake of radioactive sulfur occurred.
They noted a similar pattern of sulfate incorporation in other mucus-
secreting cells and in chondrocytes. Sulfation in the chondrocyte had been
followed by BÉLANGER (1954 b) and subsequently localized in the Golgi
apparatus by GODMAN and LANE (1964), FEWER et al. (1964), PROCKOP et al.
(1964), and others.

In 1973, YOUNG undertook a somewhat more extensive study to determine
whether sulfate metabolism is a function of a particular organelle. He noted
that different cell types have a wide distribution of sulfated compounds
containing not only protein-polysaccharides but also steroids, phenols,
thyroxine derivatives, arylamines, and lipids. He pointed out that a two-
step process is involved in formation of compounds with active sulfur groups
and raised the question of whether the enzymes involved were characteristic
of a particular organelle or occurred at different sites in different types
of cells. "During biosynthesis of these molecules, sulfate is first activated
by ATP in a two-step sequence requiring two separate enzymes. ATP-
sulfurylase catalyzes the reaction between ATP and $SO_4$ to give adenosine
5'-phosphosulfate (APS). Then APS is phosphorylated by ATP to form
3'-phosphoadenosine APS-phosphokinase. Next, the activated sulfate is
transferred to the acceptor molecule by a class of enzymes called sulfotransferases [p. 175]." YOUNG's findings indicate that in every instance
studied, the sulfate label was first identified in the membranes and vesicles

of the Golgi apparatus (Fig. 57). He concludes that although the techniques used may not retain APS and PAPS enzymes, at least sulfotransferases are associated with the Golgi apparatus membranes. GODMAN and LANE (1964) had also postulated the Golgi apparatus membrane as a site of sulfation enzyme activity. YOUNG worked with various cell types: myelocytes, reticular

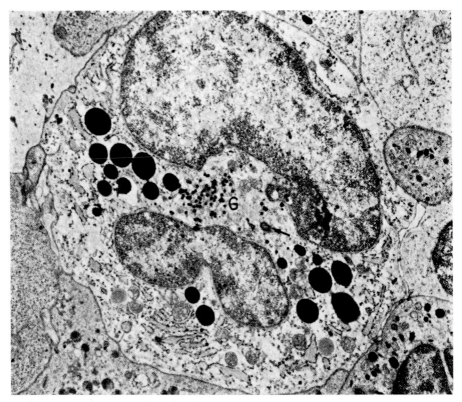

Fig. 57. Radioautograph of a basophilic myelocyte from the bone marrow of a rat 5 minutes after injection with ³⁵S-sulfate. Labelled sulfate is concentrated over the Golgi apparatus (G). From YOUNG, J. Cell Biol. **57**, 1973. Courtesy of Rockefeller University Press. ×8,100.

cells lining the sinusoids, sympathetic ganglia, endothelial cells lining small blood vessels, mast cells, Schwann cells both with myelinated and un-myelinated axons, keratocytes, fibroblasts, and ovarian follicular cells and interstitial cells (see also WEINSTOCK and YOUNG 1972 for osteoblasts and ameloblasts). These cell types are functionally quite different, yet all gave indications of sulfation of materials by the Golgi apparatus. It must, there-fore, be concluded that a wide spectrum of components is sulfated in the Golgi apparatus.

It is not possible to tell from YOUNG's micrographs whether the develop-ment of the enzymes involved in sulfation is a feature of maturation of the membranes across the apparatus nor does the resolution provided by radioautography permit concluding with safety whether this characteristic is associated with the membranes or with the secretory product. The observation does permit the localization within the apparatus of a specific step in metabolism and again one which is responsible for determining the characteristics of extracellular materials.

## G. Lipids

The synthesis and distribution of lipids is of the utmost importance in the consideration of the Golgi apparatus because substantial portions of its sometimes rapidly growing membranes are composed of lipoproteins; lipids may also be important constituents of its secretions. In considering the secretion of lipid-containing materials two cell types will be used as examples: the absorptive cell of the intestine and the liver cell. These two cell types act, in part, sequentially in the formation of the circulating lipoproteins, and studies of these cell types provide complementary information on the cellular mechanisms involved (see, for example, SHAPIRO 1967). The liver is, of course, a multifunctional organ important in carbohydrate metabolism, bile production, and secretion of plasma proteins and glycoproteins (see ERICSSON 1969, LUNDQUIST 1969). Only aspects of lipid secretion will be considered here.

Dietary fats are broken down in the digestive tract and enter the intestinal absorptive cell as monoglycerides and fatty acids which are used in the synthesis of lipids. Lipid-containing materials (lipoproteins) in the form of chylomicra are secreted laterally (Fig. 58) (CARDELL et al. 1967) from the cell and transported by way of the lymphatics to the bloodstream. Much of the lipid is removed by the liver cells (though there is also uptake by fat cells and heart muscle) where a major part of it is again used for resynthesis and secretion back into the blood. The chylomicra secreted by the liver are characterized by their content of both low and very low density lipoproteins.

The absorption of circulating lipids, the transformation of lipids within the cell, and the transport of chylomicra to and through the cell membrane have all been subjects of intense interest to the physiologist, and the early studies are sometimes confusing in their interpretations. It therefore seems in order to start with some studies that implicate the Golgi apparatus beyond any question even though there were light microscopy studies of lipid-secreting glands that suggested such implication much earlier. One early electron microscopic radioautographic study was that of STEIN and STEIN (1967 a) who introduced $^3$H-palmitate and $^3$H-glycerol into fasted and ethanol-treated rats. In the liver 2 minutes after injection, label from the lipid precursor was seen over both rough and smooth elements of the endoplasmic reticulum which the investigators designated as the site of glycerol esterification. At 5 minutes it was also seen over droplets apparently representing storage lipids. From 10 minutes on, concentrations of label were seen over the Golgi apparatus (Fig. 59). The labelled material was shown to be mostly triglyceride. This study suggests the endoplasmic reticulum as the site of triglyceride formation making this an instance in which both components of the secretion (the lipid and the protein) are formed in this organelle. The Golgi apparatus was clearly involved in the subsequent

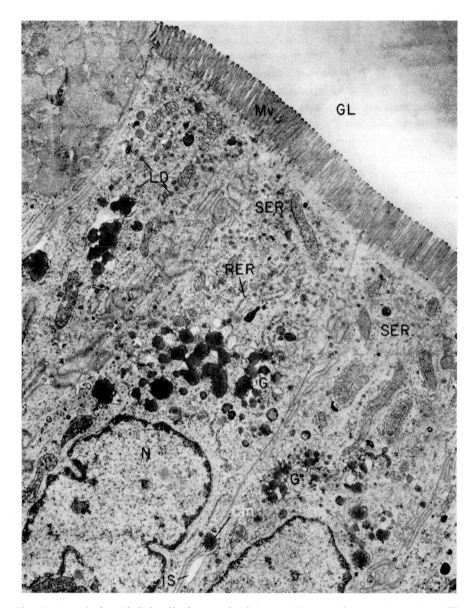

Fig. 58. Intestinal epithelial cells from a fat-fed rat. *GL* = gut lumen, *Mv* = microvilli, *LD* = lipid droplets, *SER* = smooth endoplasmic reticulum, *RER* = rough endoplasmic reticulum, *G* = Golgi apparatus, *N* = nucleus, *IS* = intercellular space, *Cm* = chylomicron. From CARDELL et al., J. Cell Biol. **34**, 1967. ×7,200.

transport of the material, and the question arose as to what specific part it played. STEIN and STEIN settle for a statement that the material is "processed" in the Golgi apparatus and the vesicles derived from this organelle act as transport vesicles by means of which the material is passed through the cell membrane into the intercellular space. JONES et al. (1967) note that some of the secreted lipoprotein also contains a carbohydrate component and suggest the Golgi apparatus may be involved in adding this component to the forming chylomicra.

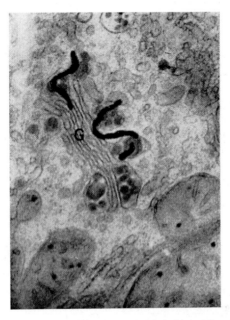

Fig. 59. Radioautograph showing concentration of label from ³H-palmitate over particles close to the Golgi apparatus (G). Section of the liver of a fasted rat. From STEIN and STEIN, J. Cell Biol. 33, 1967 a. Courtesy of Rockefeller University Press. ×34,000.

Much of the information on liver lipoprotein secretion has been summarized by CLAUDE (1970) in a discussion correlating morphology with the proposed pathway. He suggests that triglyceride synthesis may be limited to the smooth endoplasmic reticulum and that the protein and perhaps the phospholipid components are synthesized in the rough endoplasmic reticulum. Visible lipoprotein droplets are first seen just prior to their transit to the Golgi apparatus from regions involving continuities between the rough and smooth endoplasmic reticulum (Fig. 18). There is no apparent change in the lipoproteins droplets either in the Golgi apparatus or during transport to the cell surface.

The involvement of the Golgi apparatus in the secretion of lipid from the liver can be contrasted with another study by STEIN and STEIN (1967 b) in which ³H-oleic acid or ³H-palmitic acid was injected into the venous system of lactating mice. One minute after injection most of the labelled fatty acid in the gland was esterified and label was located over the rough endoplasmic reticulum or over lipid droplets. These results again suggest the transformation of fatty acids into glycerides in the rough endoplasmic reticulum, but in this case there was no subsequent movement of labelled materials to the

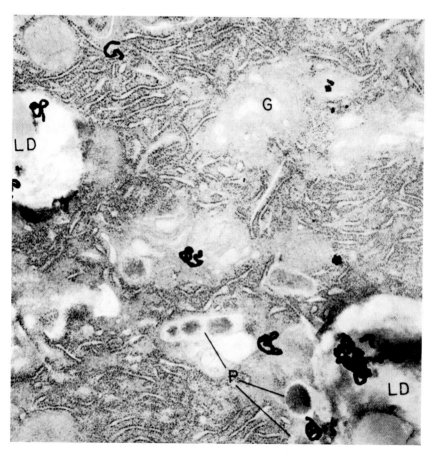

Fig. 60. Radioautograph of a portion of a cell from a mammary gland of a lactating albino mouse. Concentration of grains appear over lipid droplets (*LD*) one minute after injection of $^3$H-oleic acid. Granules (*P*) in the Golgi (*G*) vesicles are not labelled. There is no concentration of grains at any time over the Golgi apparatus. From STEIN and STEIN, J. Cell Biol. **34**, 1967 b. Courtesy of Rockefeller University Press. ×25,000.

Golgi apparatus. As is discussed in the section on milk secretion, data from various studies are consistent with the concept that lipid secretion during lactation does not involve the Golgi apparatus (Fig. 60). There seem to be different pathways by which lipids may be released from the cell as may also be the case for other materials.

In intestinal cells JERSILD (1966) had observed that when $^3$H-glucose was introduced with fatty chyme there was, after a period of time, an accumulation of labelled material in the endoplasmic reticulum and then in the Golgi apparatus. He had concluded that in absorbing intestinal cells triglyceride synthesis was localized in the endoplasmic reticulum particularly those regions close to the terminal web that underlay the modified surface of the cell. The labelling of the Golgi apparatus he looked upon as simply suggesting concentration in this organelle. DERMER (1968), in a paper that is devoted mainly to the effect of fixatives on lipids, also shows the endoplasmic reticulum to be the site of triglyceride synthesis.

CARDELL et al. (1967) made a detailed study of the absorption and modification of fats in the intestinal epithelial cell of the rat. This study confirmed earlier conclusions that fats enter the absorptive cells after being broken down into fatty acids and monoglycerides. Once in the cell they find their way into the smooth endoplasmic reticulum where synthesis of triglycerides takes place. The absorption of monoglycerides and fatty acids after corn oil-feeding of fasted rats is accompanied by a considerable modification of the smooth endoplasmic reticulum, and what appear to be aggregations of lipid can be seen within its lumen. The smooth endoplasmic reticulum is most conspicuous in the apical regions of the cell, where it is continuous with the rough endoplasmic reticulum in which lipid droplets (or chylomicra) can also be seen. With feeding there is also a modification of the Golgi apparatus with an enlargement of the Golgi *vacuoles* (secretory vesicles) which contain accumulations of lipid droplets. They note that the involvement of the Golgi apparatus is difficult to explain but that in contrast to the small droplets seen in the smooth endoplasmic reticulum, the Golgi vacuoles frequently contain large droplets which seem to result from fusion of the smaller droplets. It seemed possible, therefore, that some modification of the secretory material occurs in the Golgi apparatus.

In a later study, FRIEDMAN and CARDELL (1972 a) presented evidence that the Golgi apparatus may play a part in completing the formation of chylomicra. They also suggest that a principal factor involved was the envelopment of the lipid by a membrane so that it might be released from the cell. They found that when they treated the cells with puromycin, the formation of Golgi apparatus membrane was substantially reduced, the cell was not able to discharge chylomicra normally, and the lipids tended to accumulate in droplets sometimes (Fig. 61) nearly filling the cell. Their working hypothesis is explained in Fig. 62. For further details, see article.

FRIEDMAN and CARDELL (1972 a) also note that the vesicles derived from the endoplasmic reticulum transporting lipid to the Golgi apparatus appear to fuse only with the regions of the apparatus adjacent to the forming secretory vacuoles, *i.e.*, toward the distal face. This is consistent with the view already expressed that some degree of maturation of the cisternae within the Golgi stack may be essential for the carrying out of some of its functions. SAGE and JERSILD (1971) made a somewhat different interpretation from staining studies of the rat intestinal absorptive cell. They found that polysaccharide staining was limited to the face of the Golgi apparatus opposite that in which accumulation of lipid droplets occurred. They suggest that the organelle may be bifunctional with secretion of lipid droplets from one face and buildup and release of glycoproteins toward the other.

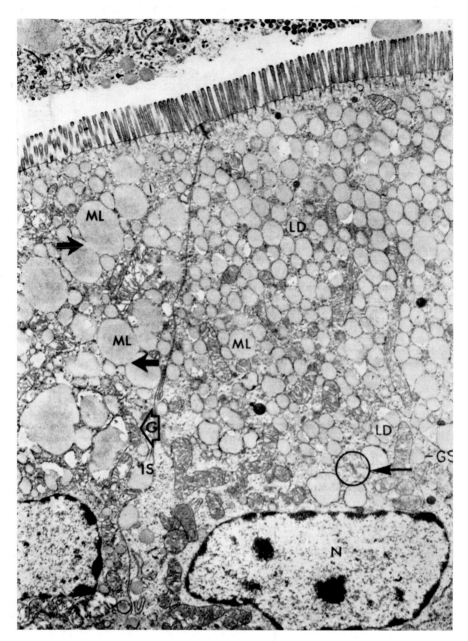

Fig. 61. Portions of two puromycin-treated intestinal absorptive cells of the rat after the administration of corn oil. The involvement of the Golgi apparatus has been blocked and there is an accumulation of lipid droplets (*LD*) and matrix droplets (*ML*) in the cytoplasm, the Golgi apparatus (*G*) and the Golgi cisternae (*GS*). From FRIEDMAN and CARDELL, J. Cell Biol. **52**, 1972 a. Courtesy of Rockefeller University Press. ×7,200.

In a more detailed study of the secretion of chylomicra from the intestinal absorptive cell, FRIEDMAN and CARDELL (1972 b) present evidence that the release of chylomicra depends upon fusing of Golgi apparatus-derived membrane and the plasma membrane of the cell. This would make the transport of chylomicra out of the cell a matter of standard secretory procedure.

Using a quite different approach, HIGGINS and BARRNETT (1971) have undertaken an ultrastructural study of the cytochemical localization of acyl-

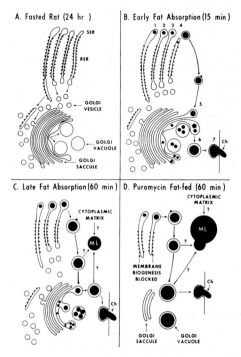

Fig. 62. Diagram of the buildup and transformation of absorbed fat by the intestinal absorptive cells of a fasted rat after fat administration. Note that transfer is from the smooth endoplasmic reticulum to the distal face of the Golgi apparatus. Exocytosis involves membrane interactions. In the pyromycin-treated material fusion does not take place and there is a formation of granules within the cell. From FRIEDMAN and CARDELL, J. Cell Biol. **52**, 1972 a. Courtesy of Rockefeller University Press.

transferases which may be involved in the formation of triglycerides in the intestinal absorptive cell. Their findings of staining product in the smooth and rough endoplasmic reticulum and in association with the outer cisternae of the Golgi apparatus are in keeping with the pathway that had been previously proposed. In a similar investigation of liver cells, however, BENES et al. (1972) and HIGGINS and BARRNETT (1972) came to a somewhat different conclusion. In liver cells the staining product occurred in both smooth and rough endoplasmic reticulum, but was not found in the Golgi apparatus. These investigators suggest the localization of acyltransferases might reflect the sites of synthesis of the lipid components of the membrane rather than

those of the secretory product. These studies point up one of the problems
in the study of lipid synthesis and secretion as some of the classes of lipids
found in chylomicra are the same as those of membrane systems (triglycerides,
phospholipids, cholesterol). This problem of course also affects the interpretation of biochemical and radioautographic studies (for example, STEIN
and STEIN 1969).

Whether by interference with the formation of membranes or of product,
interruption of the release of chylomicra from either the intestinal or liver
cells may have dire consequences for the animal. For example the congenital
disease involving a deficiency of β-lipoproteins (abetalipoproteinemia) is
accompanied by a complex of clinical manifestations. According to SCHU-
MAKER and ADAMS (1969) this condition is brought about by a homozygous
mutant allele and includes defects in the nervous system, red blood cells, and
retinal pigmentation. Basically, the condition appears to be one in which
monoglycerides and free fatty acids enter cells normally and there is a normal
conversion to triglycerides in the endoplasmic reticulum. Chylomicra are
not made, however, and the result is a fatty acid deficiency. The condition
is of interest here because it has been shown (DOBBINS 1966) that in indi-
viduals with this disease the Golgi apparatus does not function normally in
that its membranes do not become distended with lipid droplets. Although
this defect may be primarily related to the synthesis of the protein components in the rough endoplasmic reticulum, effects on the Golgi apparatus may
also be involved.

In the brief summary possible after this group of more or less restricted
studies it is possible to point out that the formation of lipoprotein going into
membrane extension or into secretory products seems to take place with
respect to both components in the endoplasmic reticulum. Nonetheless the
Golgi apparatus is apparently in some way involved. In the case of the
circulatory lipoproteins it appears to provide membrane within which these
materials are transported to selected segments of the plasma membrane for
secretion and it may contribute to those activities which change individual
lipid droplets to chylomicra.

## H. Glycolipids

Glycolipids and related macromolecules that may contain components of
all the principal classes of organic compounds have been very little studied
from the standpoint of their distribution in the cell and there appear to be no
reliable data associating them specifically with the functions of the Golgi
apparatus. Hence, any consideration of the glycolipids and the Golgi
apparatus falls into the realm of speculation. The speculation is, however, not
without some basis in an understanding of the part in cellular functioning
seemingly played by the Golgi apparatus.

OSEROFF et al. (1973) have emphasized the association of glycolipids with
the plasma membrane. It has been postulated (see DAUWALDER et al. 1972,
WHALEY et al. 1972 for discussion and references) that the plasma membrane
is basically assembled in the Golgi apparatus and components of it are then
transferred to the cell surface. It seems possible, therefore, that the apparatus

might be involved in glycolipid assembly. This possibility is reinforced by some other observations which unfortunately are no more direct. As has been pointed out by CARTER et al. (1965) the classification and nomenclature of the glycolipids is in a very confused state. The glycolipids do, however, include compounds that have as constituents galactose, fucose, and sialic acid—carbohydrate groups which have been shown (for example, in glyco-protein synthesis, see Section VI, A) to be assembled into products in the Golgi apparatus. Although considerable data suggesting the involvement of glycoproteins in surface characteristics are available, the extent to which glycolipids (either the lipid or carbohydrate portion) may contribute to specificities on the cell surface is not known. However, the introduction of "foreign" genetic material into a cell which affects many aspects of cellular metabolism including changes in glycoproteins and alterations of the cell surface also affects glycolipids, in some instances including the carbohydrate portion (see OSEROFF et al. 1973). One has to suppose that they either involve different mechanisms of assembly and transport or that the Golgi apparatus may play some part in giving them genetically controlled or influenced characteristics.

## I. The Mammary Gland

The synthesis and secretion of milk varies substantially from one species to another. It involves the correlated functioning of a number of different secretory mechanisms with the Golgi apparatus playing a key part but with other forms of secretion being quite as essential. Its initial secretion depends upon profound changes in many different tissues of the body and is subject to a complex multiplicity of stimuli (see REYNOLDS and FOLLEY 1969, COWIE and TINDAL 1971). The production of milk is a cyclic function of the alveolar epithelial cells of the mammary gland.

PATTON (1969) gives the approximate composition of cow's milk as 3.8% fat, 3.2% protein, 4.8% carbohydrate, 0.7% minerals, and 87.5% water. Being the basis of a large industry the precise characteristics of cow's milk have been somewhat better studied than those of the milk of most other mammals. It contains in addition trace amounts of many other substances. In fact, its characteristics of both composition and structure have tempted many investigators to describe it as fluid tissue.

At another extreme of the range of composition is the milk of seals and walruses which is as much as 50% lipid and has little if any lactose. PATTON (1969) has noted that as many as 150 different fatty acids can be found in milk but the principal ones are oleic acid, palmitic acid, and stearic acid along with such fatty acid chains as butyric and caproic acids. As noted previously radioautographic studies of fatty acid incorporation (STEIN and STEIN 1967 b) indicate esterification within the endoplasmic reticulum and accumulation in the fat droplets without intervention of the Golgi apparatus. The secreted lipids are nonetheless membrane-bounded. Most studies indicate that fat droplets in milk are unique in being bounded by membranes which are derived from the plasma membrane of the mammary cell during lactation (for an alternative possibility, however, see WOODING 1973). This

6*

is quite a different fate for membranes than commonly postulated in cases where membranes contribute to cell expansion or are presumably recycled. The fat droplet membrane has been much studied in relation to the characterization of different types of milk and has been shown to contain some unique enzymes as well as most of the carotene and vitamin A of the product (PATTON 1969, KEENAN *et al.* 1970 b, PATTON and TRAMS 1971).

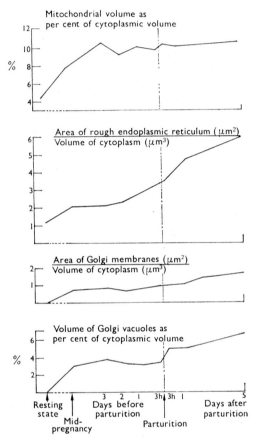

Fig. 63. Development of organelles relative to the volume of the cytoplasm during pregnancy and lactation in mice. Note that the endoplasmic reticulum and the Golgi apparatus continue to develop during the first days of lactation. From HOLLMANN, in: Lactogenesis: The initiation of milk secretion at parturition (REYNOLDS, M., and S. J. FOLLEY, eds.). Philadelphia: University of Pennsylvania Press. 1969.

Normally, the aggregation of these droplets varies considerably to give distinction in the natural products between milk and cream but most commercial milk is now homogenized, a process which stabilizes the fluid by achieving a relatively uniform size of the fat droplets. The heavy, creamlike milk of the aquatic mammals relates to the requirements that these mammals be capable of the rapid buildup of insulating materials.

The proteins of milk appear to be formed and transported along the same pathway as the proteins of other secretory products. The nonlactating alveolar cells show little differentiation. There is little rough endoplasmic reticulum and only a small inconspicuous Golgi apparatus. There are few microvilli at the apical surface (WELLINGS 1969). The single conspicuous feature is the presence of a number of large fat droplets. Following parturition (Fig. 63) there are substantial increases in the rough endoplasmic

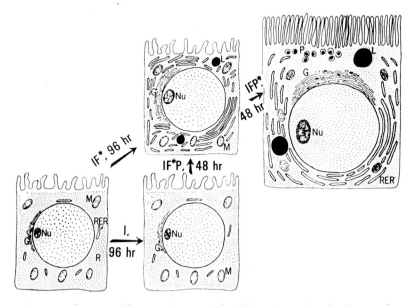

Fig. 64. Diagram of proposed hormonal control of cellular changes in what become lactating cells of the mid-pregnant mouse. $R$ = ribosomes, $G$ = Golgi apparatus, $M$ = mitochondrion, $P$ = protein granules, $RER$ = rough endoplasmic reticulum. The supposedly effective hormones are indicated by asterisks. $I$ = insulin, $F$ = cortisol, $P$ = prolactin. From MILLS and TOPPER, J. Cell Biol. **44**, 1970. Courtesy of Rockefeller University Press.

reticulum, marked hypertrophy of the Golgi apparatus, and development of many microvilli at the surface. There are some other notable cellular changes affecting the mitochondria and ultimately resulting in the release of the fat droplets. This transformation has been brought about by MILLS and TOPPER (1970) *in vitro* in explants from mammary glands of pregnant mice. The transformation depends upon the synergistic action of insulin, hydrocortisone, and prolactin (Fig. 64). When such fully differentiated cells are stimulated by oxytocin, secretion of milk into the ducts begins; its continuation depends upon its periodic removal. If it is not removed, secretion stops. Radioautographic studies (WELLINGS and PHILP 1964, ROHR et al. 1968) provide data consistent with the passage of milk proteins through the Golgi apparatus. This passage of protein granules occurs in conjunction with a 30–40 Å thickening in the membranes (HELMINEN and ERICSSON 1968 a). This appears to be another instance in which membrane is modified in the Golgi apparatus until it approximates the thickness of the plasma membrane.

The number of proteins may be high but the principal protein is casein which has some unique characteristics. There are four types of casein: alpha, beta, gamma, and kappa which again, according to PATTON, make up 50, 30, 5, and 15% of the total casein. Kappa casein is of particular interest because it contains sialic acid, presumably from what is known of other instances, added in the Golgi apparatus. Alpha, beta, and gamma caseins aggregate in the presence of calcium ions. Kappa casein resists such aggregation and up to a point acts as a protectant and keeps the milk fluid. When rennin is added to milk the sialic acid is separated from the kappa casein and it loses this protective characteristic. The result is the famous mixture of curds and whey. The curds form the basis of cheese-making. This is a useful everyday example of the manner in which the morphology of a biological substance is controlled by the presence of a carbohydrate group, presumably added in the Golgi apparatus and cleaved off extracellularly. One must also suspect some role of the Golgi apparatus in the processing of another unique protein—β-lactoglobulin. β-lactoglobulin is another substance of importance in the processing of cheese and other forms in which milk products are marketed.

The carbohydrate of milk is lactose, a sugar peculiar to milk and consisting of a galactose molecule and a glucose molecule. PATTON (1969) notes that the working out of the two enzymes involved in linking galactose and glucose represented the first illustration of a specific function of any of the milk proteins (for some of the unique features of the lactose synthetase system see GINSBURG and STADTMAN 1970). As one would suspect from the common building of galactose into glycoproteins in the Golgi apparatus there is now evidence for accumulation of lactose in that organelle (KEENAN et al. 1970 a). BREW (1969) (see also LINZELL and PEAKER 1971) has assumed that lactose synthetase is one of the component enzymes of the Golgi apparatus membranes by the time they reach the stage of development at which sections of them are pinched off to become secretory droplets. His assumption that lactose is secreted along with the proteins by the Golgi apparatus explains some aspects of lactose movement since the sugar does not appear to pass through the membrane under certain circumstances. The enzymes responsible for the synthesis of lactose also appear to be among the factors essential to the induction of lactation. However, JONES (1972) has questioned the time relationship between the conspicuous hypertrophy of the Golgi apparatus with parturition and the increase in lactose synthesis.

Investigators have speculated (PATTON 1969) that at least three mechanisms are involved in the secretion of milk. Smaller molecular components go directly through the plasma membrane; proteins and glycoproteins (and perhaps lactose) are processed through the Golgi apparatus; and the fat droplets are budded out of the cell with an enclosing membrane derived from the plasma membrane.

Being as it is the sustaining foodstuff of young mammals it is not surprising that milk is a substance of comprehensive constitution, nor is it surprising that the secretion of many compositionally different materials from the cell requires the coordinated functioning of a number of different methods of secretion. Though beyond the scope of this review an additional component of milk needs mention. At least in some animals immunoglobulins are transferred from mother to young in milk (LASCELLES 1969, WILD 1973). This process involves the selective uptake of the immunoglobulins in the digestive system of the young and their subsequent transport to the bloodstream. Furthermore, this is accomplished without degradation by the normal catabolic enzymes of the gut epithelium.

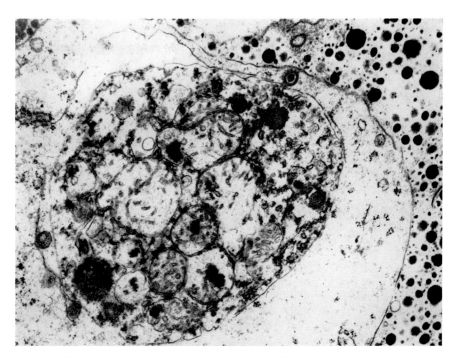

Fig. 65. Indication of the involution of a lactating cell of a rat. A large cytosegresome containing altered mitochondria. From HELMINEN and ERICSSON, J. Ultrastruct. Res. 25, 1968 b. ×16,000.

Considerable attention has been given to the involution of mammary gland cells (Fig. 65). WOESSNER (1969) has discussed the functions of lysosomal enzymes in such cells and HELMINEN et al. (1968) and HELMINEN and ERICSSON (1968 b, c, 1970 a, 1971) have followed the changes in various lysosomal enzymes including acid phosphatase, cathepsin D, arylsulfatase, acid DNase, and β-glucuronidase in involuting cells and noted that they increase with progress of the changes. They were, however, more inclined to think this a symptom of the change rather than a primary cause. WOESSNER (1969) postulated that even material in the lumen might be phagocytosed and lysed. Regardless of the mechanism, involution not only brings about a cessation of lactation but a considerable simplification of the morphology of the cell. The cell may, of course, be redifferentiated for lactation in another cycle.

The physiology of milk is touched on here only very simply. It is complex and of wide concern for the easy contamination of this multiple secretion by factors that may be deleterious and the possible effects of drugs used for other purposes on the efficiency and completeness of the process of lactation.

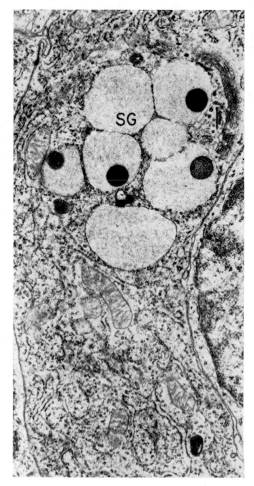

Fig. 66. Micrograph of a relatively early stage in the formation of secretion granules (*SG*) from Paneth cells of the mouse. The difference between the inner contents of the granules and the surrounding material is becoming apparent. From TROUGHTON and TRIER, J. Cell Biol. **41**, 1969. Courtesy of Rockefeller University Press. ×7,700.

## J. The Paneth Cell

Among the cells that arise in the crypts of Lieberkühn of the intestinal tract are the Paneth cells. TONER *et al.* (1971) have called the Paneth cells the most puzzling cells of the gastrointestinal tract. They occur in some species in large numbers, do not occur in other species at all, and sometimes occur only in the presence of pathological conditions.

Paneth cell renewal in the mouse has been studied by TROUGHTON and TRIER (1969). They found that compared to the goblet cell, Paneth cells are renewed and differentiated slowly and have a much longer life span in the animal. In general appearance, the cells resemble high-protein-secreting cells, but they differ in some aspects from other cells of this type (such as the pancreatic exocrine cell). They appear to secrete continuously irrespective of feeding schedules (TRIER *et al.* 1967), and in many species the mature secretory granule is composed of morphologically distinctive regions: a rela-

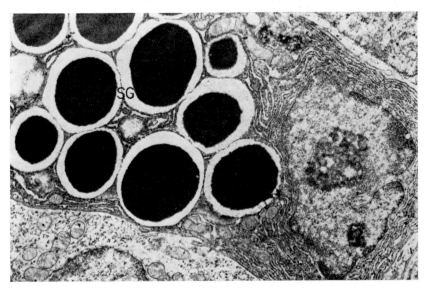

Fig. 67. Late stage in the differentiation of secretory granules (*SG*) in a Paneth cell. The granules contain a dense core surrounded by a paler mass of material. From TROUGHTON and TRIER, J. Cell Biol. **41**, 1969. Courtesy of Rockefeller University Press. ×11,500.

tively electron-dense core with a light periphery (Figs. 66 and 67). SELZMAN and LIEBELT (1962) distinguished between these regions noting that the core is protein rich and the peripheral region consists largely of acid mucopolysaccharides. These results have been confirmed by SPICER *et al.* (1967) including the basic nature of the protein cores.

Protein synthesis and transport in mouse Paneth cells was followed by TRIER *et al.* (1967) by light microscopic radioautography following the incorporation of $^3$H-leucine. Close comparison of the labelling of various regions of the cell with the ultrastructural morphology suggested the usual sequence from the endoplasmic reticulum, to the Golgi apparatus, the granules, and finally to the lumen. $^3$H-glucose incorporation was followed by HALBHUBER *et al.* (1972). Their findings are again consistent with a major portion of polysaccharide synthesis taking place in the Golgi apparatus. However, the early prosecretion granules derived from the apparatus were able to continue synthesis of polysaccharides to a minor extent. BEHNKE and MOE (1964) have studied the development of differentiating Paneth cells of the rat and have shown numerous interconnections between the nuclear envelope and the endoplasmic reticulum, calling attention to PARKS' finding (1962) that the endoplasmic reticulum may be derived developmentally from the nuclear envelope, a postulate now supported by some additional evidence (WHALEY et al. 1971). They call particular attention to the presence of structures showing periodicity in the nuclear envelope, the endoplasmic reticulum, and at least occasionally in the secretion vesicles. The significance of this apparently paracrystalline material is not known.

Many investigations of the Paneth cell have called attention to the relatively large number of lysosomes found in these cells. There is a temptation to speculate on the relationship between this consistent representation of lysosomal activity and the association of this cell type with numerous disease conditions involving modification of specific areas of the gastrointestinal tract. ERLANDSEN and CHASE (1972 a, b) present data that the Paneth cells may act as fixed phagocytes capable of taking up and digesting via lysosomal action a variety of intestinal microorganisms. They suggest that this function plus the secretion of antibacterial agents with the discharge of Paneth cell granules may provide a means of controlling the intestinal flora within the crypt lumen. It is worth noting too that the Paneth cells share with certain other high-protein-secreting cells an ability to respond to pilocarpine by the export of increased amounts of material. TRIER et al. (1967) found not only a fairly direct stimulation of export in response to pharmacological doses of pilocarpine but also a modification of cellular metabolism reflected in enhanced protein synthesis. Their micrographs show aggregations of secretion granules during exocytosis suggesting that this may be a mass response.

TONER et al. (1971) suggest that Paneth cell secretions might contribute digestive enzymes to the intestinal lumen. DE CASTRO et al. (1959) have made the interesting suggestion that the Paneth cells in the Brazilian anteater secrete a chitinase. Other investigators have made some quite different suggestions; for example, CREAMER (1967) and CREAMER and PINK (1967) have proposed that the cells secrete a lysozyme which has an antibacterial action.

The pattern of differences in regions of the secretory granules is characteristic of several types of secretion. This example does allow some implications to be drawn about the functioning of the Golgi apparatus which were not evident from the systems already discussed. It can be suggested that there may be a particular pattern of enzymes distributed spatially in the Golgi cisternae or in vesicles derived from them. Such a distribution might result in the sequestering of differing secretory products away from each other (perhaps to prevent their interaction). Thus, in addition to the differentiation or maturation of cisternal membranes across the stack, there may be discrete areas of membrane specialization within individual cisternae (see also OVTRACHT and THIÉRY 1972).

## K. Brunner's Glands

Muscosal secretion characteristic of the gastrointestinal tract is further illustrated by studies of Brunner's glands. FRIEND (1965) has presented an excellent diagram of this long-known type of cell showing a considerable development of the rough endoplasmic reticulum in the basal portion of the cell but a somewhat uncharacteristically extensive development of the Golgi apparatus in which it occupies a large portion of the cell in a supranuclear position (Fig. 68). Secretion granules are pictured in the apical portion of the cell in various stages of development. The radioautographic studies of these cells conducted by ROHR et al. (1967) and SCHMALBECK and ROHR (1967) indicate, with certain variations in time limits, similar pathways for the protein and carbohydrate components as have been described previously.

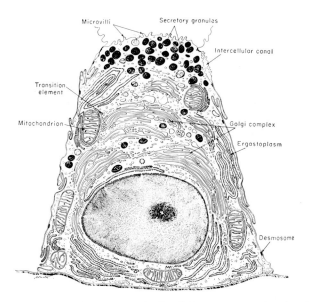

Fig. 68. A schematic diagram of Brunner's gland cell from a mouse showing the distribution of the extensive Golgi apparatus and the accumulation of secretory granules at the apex of the cell. From FRIEND, J. Cell Biol. **25**, 1965. Courtesy of Rockefeller University Press.

FRIEND (1965) has emphasized, however, that this type of cell provides an excellent illustration not only for the aggregation of material that is produced elsewhere and transported to the Golgi apparatus but also for the substantial synthetic capacity of the apparatus particularly with respect to complex carbohydrates. The morphology of the cells appears to differ significantly from one species to another as do some of the cells' activities. In most cases it is possible to demonstrate cytochemically fairly high concentrations of sulfated mucopolysaccharides. The findings indicate a transferral and perhaps the transformation of proteins, the associated assembly of carbohydrate moieties, and sulfation as successive steps involved in the formation of the secretory product. FRIEND has pointed out, however, giving credit to JENNINGS and FLOREY (1956), that the secretory products of Brunner's glands in the mouse are not sulfated. FRIEND anticipates some later work not only by calling attention to the model experimental material which cells of Brunner's glands provide but also suggesting that their secretory products may impart certain characteristics to the exposed cell surfaces.

So far, examples of materials secreted via the Golgi apparatus have included enzymes, hormones, etc.—compounds which can be looked upon as
having vital roles in the metabolic balance of cells and organisms. Secretion
products of other cell types may provide quite different functional attributes.
For example, though metabolic functions are known, many elements of the
connective tissue system are most noted for giving structural integrity to
tissues, organs, and organisms and for providing a basis for the capacity for
movement.

# VII. The Formation of Structural Components
## A. Intercellular Matrices

Cells responsible for chondrogenesis were among the favorite objects for
study in the early phases of the investigation of the Golgi apparatus. Then
when attention turned to secreting glandular cells, chondrocytes were somewhat neglected.

GODMAN and PORTER among other investigators redirected attention to
connective tissue cells (1960) by describing in detail the differentiation of
chondroblasts from relatively early stages to the point of significant secretion.
They showed that this differentiation was accompanied not only by growth
but also by extensive development of the rough endoplasmic reticulum and
the Golgi apparatus and the formation by the Golgi apparatus of large
numbers of membrane-bounded vesicles (Fig. 69). This is still one of the
best illustrations of differentiation of the components of a secretory system.
REVEL and HAY (1963) set out to explore whether or not the protein
collagen is assembled and transported in the same general manner as the
protein secreted by the exocrine pancreas cell. They patterned their methods
after those used in the radioactive-labelled-time-sequence experiments which
defined the pathway for the pancreatic cell. Because of the high concentration
of hydroxyproline in the collagen they chose $^3$H-proline as a label. Working
with *Amblystoma* limb regeneration they found the label concentrated in the
ergastoplasm 10–15 minutes after injection; over the Golgi apparatus,
30 minutes after injection; and in the cartilage matrix from 30 minutes to
an hour after injection. They found collagen fibrils to aggregate somewhat
later a short distance from the surface of the cell. Their results thus
implicated the endoplasmic reticulum, the Golgi apparatus, and vesicles
derived from the apparatus in the secretory process, followed by more
complex structuring of the materials in the intercellular matrix. Realizing
that proteins other than the precursor of collagen are also secreted by these
cells, they nonetheless suggested that the pathway of protein assembly and
secretion, probably including collagen, followed the same pattern that had
been demonstrated in the pancreatic exocrine cells. This tended to contradict
some widely held opinions that collagen fibrils are shed into the matrix by
excortication (shedding from the cytoplasmic periphery).

The evaluation of sites at which components of the intercellular matrix
of connective tissue cells are assembled is difficult for a number of reasons.
Several different types of cells secrete such material and the morphology
of the cell as well as the composition of the materials differ. For example,

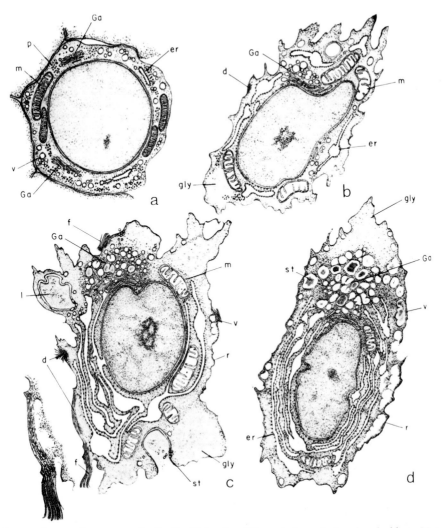

Fig. 69. Progressive stages *A—B—C—D* in the differentiation of the chondroblast. Note particularly the increase in the endoplasmic reticulum (*er*) and the Golgi apparatus (*Ga*). *v* = vesicles, *m* = mitochondrion, *p* = particles, *gly* = glycogen, *d* = dense regions, *f* = fibrils, *r* = rarefactions, *st* = stoma, *l* = pool or lake limited by double membrane. From GODMAN and PORTER, J. biophys. biochem. Cytol. **8**, 1960.

the cells often do not exhibit any polarized release of secretory materials toward a particular region of the surface making a clear tracing of materials much more difficult, and particularly in the case of fibroblasts the vesicles released from the Golgi apparatus may be relatively small complicating the problem further. Additionally, the secretions are varying mixtures of collagen precursors, chondromucoproteins, and other mucoproteins and the question of which of the components may be labelled is frequently difficult to answer. The further interactions and modification of the matrix materials after they have passed the plasma membrane also lend to the confusion.

Recognizable collagen fibrils are often not apparent close to the surfaces of the cells and become visible only after some degree of aggregation. These interactions may be particularly important. For example, FITTON JACKSON (1964, 1968, 1970), MATHEWS (1965, 1967, 1970), JACKSON and BENTLEY (1968) and others have shown that specific tissue characteristics are frequently dependent on the character of the association between the proteins and the carbohydrates in the matrix structure.

Added to these technicalities is the fact that the collagen fraction traditionally studied as a pure protein fraction has now been shown in many cases to contain carbohydrates. Carbohydrate components tend to exist in somewhat greater abundance in corneal collagen and in invertebrate collagen than in the types of collagen which have been the subject of most experiments. BOSMANN and JACKSON (1968) have reported that bovine corneal collagen contains 5.78% carbohydrate which includes certain sugar components that are known from other types of investigations to be assembled into secretory material in the Golgi apparatus. LEE and LANG (1968) have shown earthworm cuticle collagen to contain 12% galactose. Galactose is often added to other assembling secretory products in the Golgi apparatus and that organelle is apparently the principal locus of galactosyltransferase (FLEISCHER and FLEISCHER 1971). Some other matrix collagens contain even more carbohydrate.

The matrix also contains chondroitin sulfates a, b, and c, keratosulfate, heparin sulfate, and some acid mucopolysaccharides including hyaluronic acid and some sulfate esters (BRIMACOMBE and WEBBER 1964, MEYER 1969, MUIR 1969). There are obviously relationships among these components which are essential to the development of the fibrillar form of collagen. The exact nature of these relationships is obscure; some of them may depend on the interactions of the carbohydrate and protein components, indeed even the carbohydrate portion of the collagen molecules might be involved; others appear to depend upon the precise nature of the substrate on which the collagen is arranged.

A series of experiments casting some doubt on the involvement of the Golgi apparatus in the secretion of the collagenous component of the matrix was performed by Ross and BENDITT (1962, 1964, 1965, see also Ross 1968, 1969). They followed the movement of $^3$H-proline in normal and scorbutic guinea pigs during wound healing. In both groups the $^3$H-proline was maximally concentrated in the fibroblasts 4 hours after incorporation and subsequently appeared in the matrix.

In the normal animals the matrix is in part fibrillar and the banding typical of collagen is readily seen. In the scorbutic animals the matrix appears to be noncollagenous and there are other patterns in the cell ultrastructure suggesting abnormal synthesis. One of these is the existence of aberrations in ribosomal patterns in the scorbutic animals (Ross and BENDITT 1964). These observations led the investigators to conclude that $^3$H-proline might follow the same route, i.e., via the Golgi apparatus whether or not it was destined to be incorporated into normally banded collagen, suggesting some other protein component was being traced by these techniques.

These findings and the aberrations in the ribosomal patterns led Ross and BENDITT to postulate that the endoplasmic reticulum might play some part not only in the synthesis of collagen but also its secretion. They suggested that under certain abnormal conditions, stages in collagen formation may not proceed to the point where the material becomes banded. The emphasis of these investigators on the endoplasmic reticulum finds some interesting extensions in further work of theirs and other investigators.

In another experiment Ross and BENDITT (1965) introduced $^3$H-proline into normal animals and then attempted to follow it quantitatively from one site to another in a time sequence. In this instance label over the endoplasmic reticulum appeared most concentrated after 15 minutes and did not reach its maximum concentration in the Golgi apparatus until 60 minutes after incorporation. Further quantification of the amount of labelling per unit area of the cell occupied by the endoplasmic reticulum and the Golgi apparatus showed that quite unlike the behaviour of some protein-labelled components, the decrease in the label over the Golgi apparatus occurred concurrently with or even in part prior to decrease in the label over the endoplasmic reticulum. In time the label appeared in the matrix, and this pattern suggested that there was not a transfer of all the labelled material from the endoplasmic reticulum to the Golgi apparatus prior to its secretion. They came to the conclusion that some protein might be secreted directly from the endoplasmic reticulum and suggested that this might be the collagen fraction. This experiment is somewhat complicated by the general recognition that in many instances labels indicate both secretory and sedentary (non-secreted) protein (LEBLOND 1965). Ross and BENDITT's conclusion was also supported by ROHR (1965) who indicates some secretion from the endoplasmic reticulum of the synthesized protein which becomes the banded collagen without the involvement of the Golgi apparatus.

Other investigators (SALPETER 1968, COOPER and PROCKOP 1968) have suggested from experiments in which the degree of resolution is not altogether clear that a collagenous precursor may find a concentration and part of its development in the cytoplasmic matrix. To an extent at least this would indicate a partial return to the earlier theory of ecdysis.

There seemed to be general agreement that some portion of the protein of the connective tissue matrix moved along the pathway indicated for other protein-secreting cell types and was transported via the Golgi apparatus to the cell surface. For the collagen component, however, plausible evidence had been presented for three distinct secretory mechanisms: via the Golgi apparatus, via the endoplasmic reticulum, or through the cytoplasmic matrix. The differing cell types studied and the differing approaches make a consistent interpretation of these data difficult. There is, however, increasing evidence from studies of corneal cells and odontoblasts that collagen is assembled in the Golgi apparatus and transported in Golgi vesicles.

In a series of studies of avian corneal development (HAY and REVEL 1969, TRELSTAD 1970, 1971) it has been shown that the position of the Golgi apparatus appears to change in relation to the direction of secretion and that Golgi apparatus vesicles may contain fibrillar elements with a pattern of collagenlike striations. These and other data supporting the collagenous nature of the material processed through the Golgi apparatus have been reviewed by Hay and Dodson (1973). Some of the data on the interruption of collagen secretion by inhibitors also show effects on the Golgi apparatus. Coulombre and Coulombre (1972) found that L-azetidine-2-carboxylic acid (LACA) caused a simultaneous change in the Golgi apparatus and cessation of collagen secretion. Seegmiller et al. (1972 b) treated embryos with 6-aminonicotinamide (6-AN) and brought about reduction of the Golgi apparatus and reduction in the amount of fibrillar collagen (see Section XII).

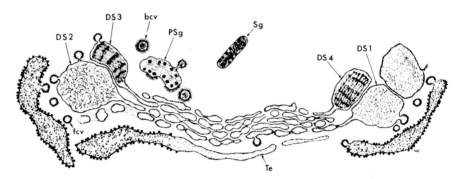

Fig. 70. Diagram of a portion of the Golgi apparatus in the odontoblast and associated structures presumably involved in the synthesis of procollagen. Collagen precursors are transported to the Golgi cisternae by transitional elements (*Te*) or by fuzz-coated intermediate vesicles (*fcv*). These are sequestered into the Golgi cisternae (*DS* 1, *DS* 2). Later they form threads (*DS* 3, *DS* 4). With further development prosecretory granules (*PSg*) are formed. Bristle-coated granules (*bcv*) may bud off from these. Condensation of procollagen follows in secretory granules (*Sg*) which migrate to the surface. From Weinstock and Leblond, J. Cell Biol. **60**, 1974. Courtesy of Rockefeller University Press.

Perhaps one of the most striking demonstrations has come from studies on odontoblasts. The relatively high percentage of collagen in the dentin matrix and the highly polarized morphology of this cell type have made for a clearer tracing of the secretory pathways. M. Weinstock and Leblond (1974) (Fig. 70) have diagrammed the character of the Golgi apparatus in the odontoblast and summarized the data for the processing of collagen precursors from early steps in the endoplasmic reticulum to the formation of oriented threadlike structures within the apparatus. They suggest that processes in the Golgi apparatus are responsible for the reorganization of disoriented entangled fibrils into the stacks of fibrils assembled in parallel arrays which can be seen in the maturing secretory vesicles.

With respect to the carbohydrate-containing components of the matrix, a number of studies implicate the Golgi apparatus as playing a primary role. Barland et al. (1968) have shown $^3$H-6-glucosamine to be a relatively specific precursor of $^3$H-hyaluronic acid. In nature, hyaluronic acid is loosely associated with protein though the association is easily destroyed. Hyaluronic acid is one of the common polysaccharides secreted by synovial cells. Barland and his co-workers found that when human synovial cells were cultured in the absence of glucose the uptake of $^3$H-glucosamine was materially enhanced. In studying the time sequence they found the early concentration of the label at 5 minutes to be over the Golgi apparatus. There was subsequently a reduction of this label concurrent with the appearance of label in the synovial fluid. They postulated a synthesis in the Golgi apparatus along with some storage here. They and other investigators (STOOLMILLER and DORFMAN 1967) have demonstrated the continued synthesis of hyaluronic acid when protein synthesis is blocked.

HORWITZ and DORFMAN (1968) attempted to provide an answer to the site of polysaccharide synthesis by an approach in which they assayed the amount of xylosyltransferase and galactosyltransferase in fractions of rough and smooth microsomes from chick cartilage. They found both enzymes to be present in the highest concentrations in the rough microsomes—a finding that is at odds with some others, at least with respect to galactosyltransferase (FLEISCHER and FLEISCHER 1971). The more extensive findings linking the Golgi apparatus to the synthesis of matrix carbohydrates have been reviewed for cartilage cells by REVEL (1970) and for ameloblasts and odontoblasts by A. WEINSTOCK and LEBLOND (1971) and A. WEINSTOCK (1972). The sulfation of matrix components in the Golgi apparatus has already been noted.

The pathway of secretion of the various components appears to be in keeping with that already described. Whether the model for collagen secretion provided by study of the odontoblast can be generalized to other collagen-secreting cell types remains to be seen. Patterns of collagen-containing matrix secretions may differ from one cell type to another, and the collagens and associated compounds are characteristic of different sorts of tissues as well as various evolutionary levels. There may well be substantial significance in the fact that M. WEINSTOCK and LEBLOND (1974) have seen fibrillar components in the cisternae of Golgi apparatus whereas some other investigators have failed to detect developments this advanced at this site. There are other examples referred to throughout this work suggesting that the stage of advancement to which the membranes of the Golgi apparatus may develop may differ. The fact that fibrillar units are at least sometimes seen in cisternae of the Golgi apparatus or vesicles derived from them might indicate either a specific spatial distribution of enzymes within this structure or some capacity for self-assembly, perhaps influenced by the associated polysaccharides as has been suggested for the intercellular matrix (see MATHEWS 1970, TRELSTAD and COULOMBRE 1971, SEEGMILLER et al. 1971).

Such interactions may be of particular importance in embryonic differentiation. HAY and REVEL (1969) and HAY and DODSON (1973) have raised the question of whether or not certain epithelial matrices may not act to induce particular developments in embryonic animals. BERNFIELD and WESSELLS (1970), SEEGMILLER et al. (1971), and BERNFIELD et al. (1972) have drawn particular attention to the possible roles of the carbohydrate components of the matrix in such processes. KOSHER et al. (1973) have shown that some aspects of chondrogenesis seemingly involved in differentiation can be enhanced in vitro by the addition of exogenous chondromucoprotein.

They suppose that processes mediated by the cell surface may be involved. The question of embryonic induction is a key one in morphogenesis. It may involve specificities genetically controlled and mediated intracellularly by the Golgi apparatus and it certainly involves extracellular remodeling of a complex system.

The range of matrix materials from seemingly amorphous products to relatively highly oriented ones again emphasizes the versatility of cells. This versatility depends in part on the development of an interrelated system of specialized cellular membranes. Thus membranes are of primary concern and appear to be the structures through which the genetic determination of the end products takes place no matter how much modification of these products there may be after they are discharged from the cell.

## B. The Plant Cell Wall

Except in a few instances the cells of plants are surrounded by walls. An examination of these walls shows, however, that both structurally and ontogenetically they bear much resemblance to the connective tissue materials that surround both cells and tissues in animals. Plant walls have a fibrillar component which makes up a latticework which comprises a fraction of the wall. The wall also contains variously associated materials including predominantly polysaccharides but also some lipids and proteins. Though principally a structural material, there is physiological evidence of a good deal of biochemical activity in the wall (for example NEVINS et al. 1967, 1968, ALBERSHEIM et al. 1969, KNOX 1973, DE NATTANCOURT et al. 1973). In the majority of plants the fibrillar component is the polysaccharide cellulose which is transported in some plant cells at least in association with other materials via a process that involves the Golgi apparatus. As is the case with the collagenous component of the animal connective tissue matrix, the extent to which this generally occurs is not yet clear. Some of the enzymes known to be characteristic of the Golgi apparatus apparently function in one or another of the processes which take place in wall formation. A few plants have walls in which the structural component is chitin or mixtures of cellulose and chitin. The cellular components involved in the formation and secretion of chitin in plants have been little explored. Structurally, the fibrillar component of the plant cell wall has a specifically modelled architecture, and one has to assume that if exocytosis of the various components of this wall is involved there must be much extracellular modelling to give the characteristic wall structure. This is particularly true since many plant cell walls are laid down while the cells are in a process of growth and many of the features are apparently determined by a combination of secretion and elongation.

Although many of the details of wall formation are not yet known the primary concern here is with the possible involvement of the Golgi apparatus. There seems to be ample proof that much of the matrix polysaccharide is formed in the Golgi apparatus and transferred to the new wall. As in the case of the animal materials there is somewhat more question about the extent to which cellulose synthesis and fibrillogenesis may or may not proceed before the material actually is in the wall. Much of the information on the secretion and characterization of plant cell walls has been summarized by NORTHCOTE (1968, 1969 a, b, 1971 a), and he has commented somewhat further on the particular roles of the Golgi apparatus (NORTHCOTE 1971 b).

Some aspects of the functioning of the Golgi apparatus in the formation of the new plant cell wall lend a unique emphasis to what now seems to be

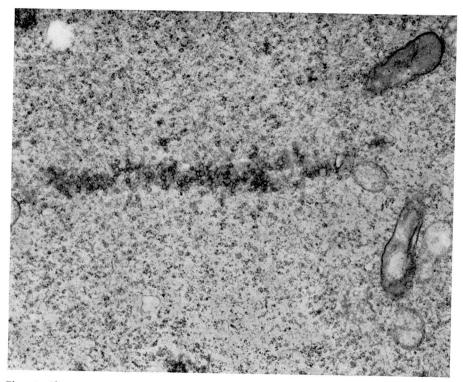

Fig. 71. Clusters of small membrane-bounded vesicles from the Golgi apparatus in anaphase of a *Zea mays* root apex cell. From WHALEY *et al.*, J. Ultrastruct. Res. **15**, 1966. ×19,000.

among the more important of its activities. Where it has been studied in detail during cell division in root tips, it has been shown that the Golgi apparatus is apparently barred from the spindle figure, but that in one of the patterns described by some of the light microscopists as well as in some electron microscopy individual Golgi apparatus take up positions seemingly without particular polarization around the presumptive spindle region in the middle of the cell. Some time during anaphase or perhaps earlier, these Golgi apparatus produce large numbers of small membrane-bounded vesicles which apparently invade the spindle region after its definition at or near its poles. These vesicles move in some sort of association with the spindle fibers toward the equatorial region of the cell where clusters of them aggregate to form groups around the centralmost spindle fibers (Fig. 71). In *Zea mays* (WHALEY *et al.* 1966) at least there is at some stage a change in the order of magnitude of the membrane-bounded vesicles and significantly larger ones are produced. These also move along a similar path to become associated with the smaller vesicles (Fig. 72). The islands of vesicles increase and after the arrival of the larger vesicles fusion begins (Fig. 73). That the basic process of cell plate

Fig. 72. Large and small vesicles in an anaphase-telophase cell of a root apex of *Spinacia oleracea*. A later stage than Fig. 71. Note that spindle fibers still show. From WHALEY *et al.*, J. Ultrastruct. Res. **15**, 1966. ×33,000.

formation involves the Golgi apparatus had been suggested earlier by WHALEY and MOLLENHAUER (1963) and has been confirmed by FREY-WYSSLING *et al.* (1964) and, except for the differences in the magnitude of the vesicles produced, by CRONSHAW and ESAU (1968), ROBERTS and NORTHCOTE (1970) and others. The movement of the membrane-bounded vesicles from the polar regions toward the equatorial regions accords precisely with what BAJER and ALLEN (1966) describe as the movement of swellings along the spindle fibers in division in *Haemanthus* endosperm in which they have studied division in detail (Fig. 74). This is an important observation inasmuch as BAJER's studies were made by light microscopy with Nomarski optics and not by electron microscopy. For additional information see the review of BAJER and MOLÈ-BAJER (1972).

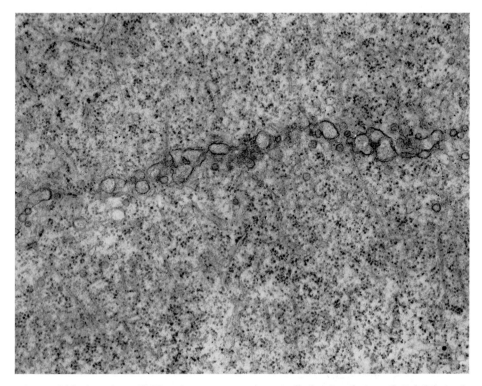

Fig. 73. Telophase in a dividing *Zea mays* root apex cell showing fusion of Golgi-derived vesicles. From WHALEY *et al.*, J. Ultrastruct. Res. **15**, 1966. ×27,000.

The fact that most plant cells divide by cell plate formation and not by cleavage or furrowing presents an unusual opportunity to determine the actual contribution of the Golgi apparatus. The process starts in the mid-region of the cell, and until very late there is no contact with existing membranes. There seems to be no question in this instance of the formation of plasma membrane by the extension of existing ones. The plasma membranes of the cells resulting from the division are contributed by the Golgi apparatus as are the substances—mostly pectins (see DAUWALDER and WHALEY 1974) and hemicelluloses—that separate them. Often the endoplasmic reticulum is abundant in the cell plate region (Fig. 75) and this has led to the suggestion that some of the materials of the plate may be derived directly from the endoplasmic reticulum (HEPLER and JACKSON 1968). The plate-formation process can be prevented by colchicine (see PICKETT-HEAPS 1967) and caffeine thus suggesting that this transfer of membrane and bounded material is a co-ordinated cellular process. Whereas the block in cell division brought about

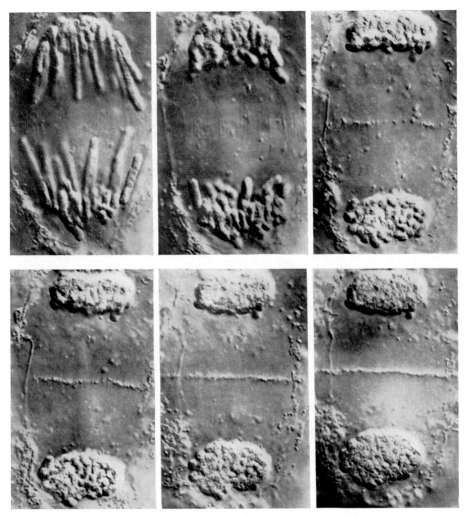

Fig. 74. Progressive stages in cell plate formation in *Haemanthus* endosperm as shown by Nomarski optics. From Bajer and Allen, J. Cell Sci. **1**, 1966. Courtesy of Cambridge University Press. ×1,140.

by colchicine treatment affects primarily microtubular arrangements and the formation of a normal spindle apparatus, Paul and Goff (1973) have suggested that caffeine treatment might more directly affect the membranes of the plate vesicles by altering some mechanism involved either in vesicle recognition or fusion. It is but one of a series of illustrations in which the Golgi apparatus appears to contribute to the plasma membrane (see also Northcote and Lewis 1968), but it is a particularly good one because it happens in a situation in which the contribution of membrane from other existing plasma membranes is initially impossible.

There is little question either about the organelle's contribution of the early wall material made up in this instance largely of carbohydrates but

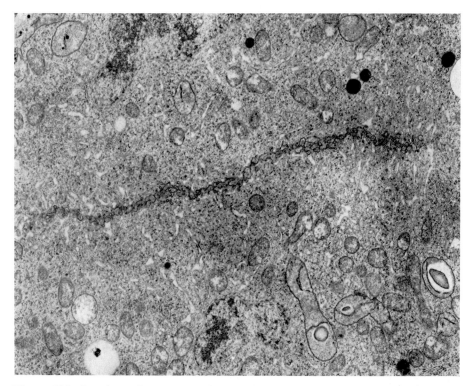

Fig. 75. Telophase in a 3-hours puromycin-treated *Zea mays* root apex cell. Plate formation appears normal but note endoplasmic reticulum in regions of the plate. From WHALEY *et al.*, J. Ultrastruct. Res. **15**, 1966. ×9,200.

also of sufficient other compounds to condition substantial further development of a complex structure. The principal question, as in the case of the development of collagenous structure in connective tissues, is the relation of the sites of particular steps in the synthesis of cellulose to the development of the final fibrillar material. This question has occupied botanists with theories in the absence of facts for a long time and led to some widely divergent points of view. Much of the conflict historically has dealt with whether the cellulose component is formed extracellularly or intracellularly and will not be reviewed here. The focus will be primarily on the question of the possible involvement of the Golgi apparatus.

First, however, attention needs to be called to the fact that except for tracing of materials that fall into the general category of pectins and hemicelluloses there has been little morphological investigation of the other specific wall components, and at this point it seems only logical to assume that some of them are synthesized and transported by the usual routes. One fairly recent finding deserves mention. It is that most plant cell walls are fairly rich in hydroxyproline (see LAMPORT 1970) which is also a major component of animal connective tissues. This makes for a puzzling situation inasmuch as $^3$H-proline introduced into plant cells has not yet been clearly followed by labelling procedures into the wall. (For various views on this controversy see ISRAEL *et al.* 1968, ROBERTS and NORTHCOTE 1972, STEWARD *et al.* 1974.)

NORTHCOTE (1968), based in part on some earlier studies by NORTHCOTE and PICKETT-HEAPS (1966), has postulated a scheme whereby carbohydrate entering the cell might move along any of several routes (Fig. 76). According to this scheme glucose entering the cell becomes part of a pool of hexose

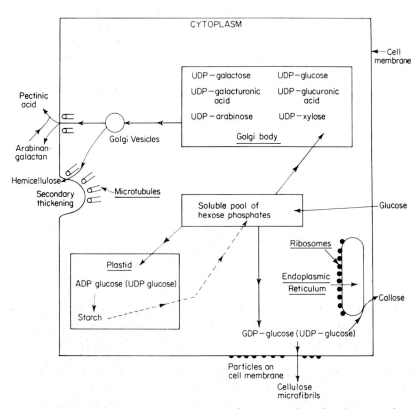

Fig. 76. Diagram representing possible pathways of various polysaccharides in a plant cell. From NORTHCOTE, in: Plant cell organelles (PRIDHAM, B. J., ed.). London-New York: Academic Press. 1968.

phosphates. If the cell is in a storage phase it may reversibly incorporate the hexose into starch in the plastids. Some portion of it makes its way directly into the Golgi apparatus where any of a number of interconversions takes place. The general references to such interconversions indicate that the Golgi apparatus is a center for a number of enzymes involved in different reactions. By producing membrane-bounded vesicles the Golgi apparatus becomes the source of the early cell wall materials, pectic acids, and hemicelluloses, which are guided into various regions of the wall by arrangements of microtubules. In this scheme glucose moves directly out through the plasma membrane. NORTHCOTE postulates that particles on the membrane with enzymatic capacities may perhaps participate in the formation of fibrils after moving away from the membrane. This represents the secretion of a precursor of the cellulose portion of the wall not involving the Golgi apparatus, completion of the cellulose molecule, perhaps in association with the plasma membrane, and fibrillogenesis in the intercellular space under the influence

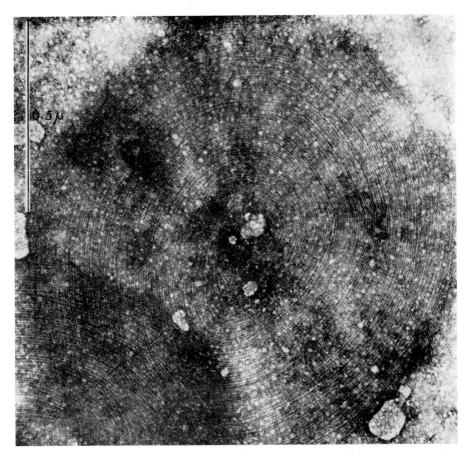

Fig. 77. A scale from the cell wall of an alga, *Pleurochrysis scherffelii*, after an extraction which presumably removes all but the cellulose fraction. From BROWN *et al.*, J. Cell Biol. **45**, 1970. Courtesy of Rockefeller University Press. ×100,000.

of other matrix materials which were synthesized and transported via the Golgi apparatus. More recently the accumulating evidence linking the Golgi apparatus to the plasma membrane has led to the suggestion that enzymes responsible for the synthesis of cellulose at the plasma membrane might be transported to that site via Golgi apparatus vesicles, giving the apparatus at least an indirect role in cellulose synthesis (see NORTHCOTE 1972).

A totally different explanation has stemmed from the studies first carried out by MANTON (MANTON and PARKE 1962, MANTON *et al.* 1965, MANTON

1966 a, b, 1967 b) of scales or crystals produced by certain algae. BROWN and his colleagues (BROWN 1969, BROWN et al. 1969, 1970) have subjected such crystals to detailed analysis and concluded that they contain cellulose, other polysaccharides, and associated amorphous components. They have depicted stages during the progressive extraction of materials which finally show a coiled fibril intact in the scales. Actually such fibrils are detectable while they are still in association with the Golgi apparatus (Fig. 77). By the large number of tests they applied this microfibril appeared to be cellulose.

The postulate put forth by BROWN et al. (1973) has applied some of the concepts developed in the study of animal cells to scale formation with the transfer of polypeptides from the rough endoplasmic reticulum to the Golgi apparatus and the exertion by these peptides of some control over the formation of the cellulose microfibrils, though they have allowed the possibility that control may also be exerted by acidic polysaccharides as suggested much earlier by STEWARD and MÜHLETHALER (1953). Regardless of the validity of this concept as applied to the development of plant cell walls, it assigns the synthesis of at least a portion of the polysaccharide groups to the Golgi apparatus which is consistent with numerous other findings.

BROWN et al. (1973) propose the use of this model for the consideration of wall building in general. It would seem that the analysis must be carried further before this is possible, though it is in order to point out that, as in the case of the study of the odontoblast by WEINSTOCK and LEBLOND (1974) fibrillar components are sometimes found within the cisternae and the vesicles of the Golgi apparatus. It may be of some consequence in regard to the extent to which structural components become fibrillar to recognize that the membrane-bounded vesicles derived from the Golgi apparatus represent components of DE DUVE's exoplasmic space (see DE DUVE 1969). Even before there is any fusion of the Golgi vesicles with the plasma membrane, the contents of these vesicles are isolated from the cell cytoplasm by a membrane. With fusion the Golgi vesicle space becomes continuous with the extracellular space and sometimes makes for substantial increases in such space. Cytochemically or by other means the contents of the Golgi apparatus vesicles can often be shown to be rich in an assortment of enzymes which may influence the character of the contained materials either within the vesicle in the cytoplasm or outside the cell. There may be great differences in the Golgi apparatus at various evolutionary levels or in tissue cells of different specializations, and products may proceed to a quite different degree of development either within Golgi vesicles or in the exoplasmic space. It is not difficult to envision, for example, that within the cell sometimes structural entities such as fibrils may be fairly fully developed in this space whereas under other conditions they may be developed only to some precursor stage. Although with regard to cellulose synthesis many questions remain unanswered in many organisms the Golgi apparatus does seem to be the site of synthesis and addition of certain of the carbohydrate moieties. In the plants these come to make up the bulk of the wall material, for in addition to the laying down of the initial wall, Golgi apparatus vesicles can be followed into the wall at much later stages (Fig. 78) (see VIAN and ROLAND 1972). They also play a part in the laying down of some of the secondary wall materials (Fig. 79) and contribute to slimes and other external material (see SCHNEPF 1969, HYDE 1970, ROUGIER 1971). It is important to recognize, however, that prokaryotic cells without Golgi apparatus often form walls of substantial

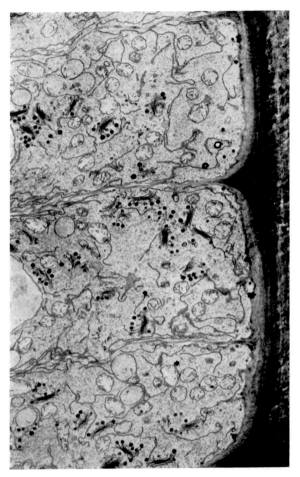

Fig. 78. Epidermal cells of a *Zea mays* root apex showing the continuing addition of membrane-bounded vesicles to the wall. From H. H. MOLLENHAUER, Cell Research Institute, University of Texas at Austin. ×5,500.

complexity. In the case of *Acetobacter xylinum* cellulose fibrils are produced in sufficient amounts to have provided a satisfactory experimental system. The synthetic mechanism appears to involve lipid-containing intermediates (glycolipids) as does cellulose synthesis in higher plants (see KJOSBAKKEN and COLVIN 1973, ELBEIN and FORSEE 1973), however, in contrast to the previously noted hypotheses bacterial cellulose can be produced in an appropriate medium not containing any bacteria, provided a fibril is introduced into the medium. Not only may the process take place extracellularly

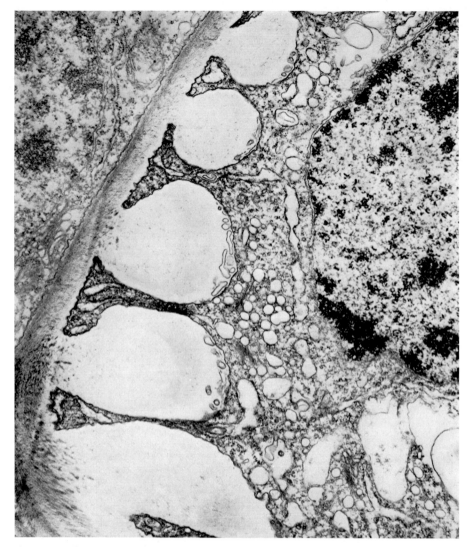

Fig. 79. Golgi apparatus contribution to secondary wall thickenings in the xylem of *Malpighia glabra*. From H. J. Arnott, Cell Research Institute, University of Texas at Austin. ×22,000.

but it may incorporate exogenous material into the fibril building (Ben-Hayyim and Ohad 1965). These differences may be related to the fact that bacteria often secrete into the extracellular medium enzymes which may, in more complex organisms, act intracellularly or be more tightly bound to the plasma membranes.

To a considerable extent the integrity of multicellular organisms is dependent upon a sequence of cellular and intercellular events in wall production that involves the Golgi apparatus. One must, however, avoid the conclusion that the wall is an exclusive concern in the determination of this integrity.

Depending upon the tissue type, cells are held together or interconnected by a wide range of mechanisms extending from desmosomes and tight junctions in animal cells to plasmadesmata in plant cells. Geometric arrangements play a part of importance in nearly all instances. One must also avoid the conclusion that the sort of frank structural components discussed here are the only constituents whose formation involves the Golgi apparatus in the maintenance of this integrity. Various bioadhesives seem often to be of concern not only in the determination and maintenance of form but also in characterizing the organism. In some sorts of organisms substances produced by the Golgi apparatus may serve to anchor the organism and act as bioadhesives between the cells and extraneous objects (for example, PHILPOTT et al. 1966, PERRY and WADDINGTON 1966, WALKER 1970, AKAI 1971, TAMARIN and KELLER 1972). In others, seemingly related substances produced by special glands or cells act as slimes which contribute to the motility of the organism (for example, HARRISON and PHILPOTT 1966, WONDRAK 1967, HARRISON 1968, PORTER 1969, ZYLSTRA 1972, ENGSTER and BROWN 1972, SOUZA SANTOS and SILVA SASSO 1973). Slimes may, of course, also serve to protect the organisms and to prevent drying.

## VIII. Coordination

### A. The Development and Functioning of the Nervous System

The Golgi apparatus was discovered in cells of the nervous system but attention turned quickly to other types of epithelial cells that were simpler and in which, as discussed elsewhere in this work, it finally became clear that the organelle acted as a focal point for a variety of secretions. The passage of substances controlling cellular development and affecting the functioning of the nervous system was pursued largely by physiological studies, partly because of the technicalities involved in the functioning of this system, and partly because of the substantial diversity of theories concerning the operation of the system. Interest in the role of the Golgi apparatus in cells of the nervous system, however, has been renewed by the growing pervasiveness of the concept that the steps in the secretion process are part of the normal metabolism of essentially all cells, and it is being strengthened by the growing awareness of the importance of the transfer of specific materials to the cell surface both developmentally and functionally.

The sequence of views leading to a common definition is summed up in a succession of well-chosen quotations used by A. D. SMITH in the concluding paper of a meeting on synaptic transmission, the proceedings of which were published in the *Philosophical Transactions of the Royal Society of London* (1971).

"Believing that the nervous system is something more than a mere system of conducting paths, I formed the hypothesis that nerve cells are true secreting cells, and act upon one another and upon the cells of other organs by the passage of a chemical substance of the nature of a ferment or proferment (SCOTT 1905)."

"The most striking morphological feature of the neuron is the tremendous accumulation within its cytoplasm of small granules associated with a well developed endoplasmic reticulum. The same type of association is found in the ergastoplasm of glandular cells ... cells which sustain an intense protein production. In the nerve this activity is implicit in chromatolysis and to certain types of generalized stress. The rapid regeneration of axons and the peculiar damming up of axoplasm proximal to a ligature are also reflexions of a continuous and rapid protein synthesis in the perikaryon ... The fact that the structure of the Nissl substance is the same as that of the ergastoplasm in glandular cells means that future analysis ... in such readily available cells as those of the pancreas and liver can be profitably applied to the nerve cell (PALAY and PALADE 1955)."

"Neurosecretion should not perhaps be used as a term to describe only the histochemically demonstrable secretory processes of nerves such as those of the hypothalamo-pituitary system. The analogies are sufficient to indicate that similar processes are involved in the production, transport and secretion of acetylcholine from other nerve endings. So these, too, may also be called neurosecretory nerves (HEBB 1959)."

"All neurons have a secretory function by which active substances are synthesized and released. Secretion may act over a short distance on specific chemical receptors or on distant receptors by way of the blood stream. Intermediary examples are the adrenergic neuro-effectors ending on smooth muscle. Neurosecretion may be produced all along the neuron or at the nerve endings. In all cases, it is stored within a membrane in vesicles which represent multimolecular quantal units of neurosecretion (DE ROBERTIS 1964)."

What emerges from these quotations is the slow modification of a number of theories of the passage of transmitter substances through neuron membranes to an acceptance of the view that these substances are formed, transported, and secreted by processes comparable to those in other types of cells. Not only does this appear to be the case with neurotransmitter substances, it also appears to pertain in instances in which substances are assembled and secreted to play a part in the development of the nervous system. It indicates a concern with secretion in all cells of the developing and functioning system and not just with certain cells that were recognized long ago as being frankly secretory (see BARGMANN 1966, STUTINSKY 1967, HOFER 1968). It has been suggested by BRUNNGRABER (1969 a) that the absence of certain specific glycoproteins or families of glycoproteins from the surface of developing nerve cells may be associated with certain deficiencies in nervous system maturation and subsequent function. BRUNNGRABER (1969 b, 1972 a, b) has developed this concept further with particular attention to the importance of the carbohydrate moieties and draws attention to the possible role of glycoproteins in normal nerve transmission (perhaps including processes such as learning) and in certain neuropathologies. A particularly interesting example of the effects of abnormal developmental control in the embryonic nervous system is seen in the genetic mutation resulting in *reeler* mice. The findings (SIDMAN 1970, 1972) suggest that in the formation of the cerebellar cortex the proper alignment of cells with normal polarity is not a simple aggregation phenomenon. There must be factors at the cell surface acting within a given time period in the developmental sequence that control not only the side-by-side alignment of like cells but also their polarized end-to-end associations which lead to the layering of discrete cell types in the mature cerebellum. As SIDMAN points out this seems to be a cell recognition problem, but recognition of what, on what part of which cells, and when? ALTMAN (1971) has presented evidence suggesting that vesicles from the Golgi apparatus do indeed contribute to plasma membrane formation in the normal development of the rat cerebellum. By analogy with other studies of cell recognition, it will be of interest to find the extent to which surface glycoproteins may be involved in developmental control of the nervous system (see DI BENEDETTA and CIOFFI 1972).

That the basic story is comparable to that in other cells is illustrated by data from experiments by DROZ (1965, 1967 a, 1969). Using $^3$H-leucine and $^3$H-arginine, DROZ and his colleagues have demonstrated the synthesis of proteins in the perikaryon. A few minutes after injection there is a high concentration of label over the Nissl substance which is rich in ribosomes. Within 20–30 minutes the highest concentration appears over the Golgi apparatus. Here as in other instances there is a separation of sedentary and migratory proteins, and other investigators (see DROZ and KOENIG 1969) have been able to follow the latter in part down the axon to accumulation at the nerve endings.

DROZ (1967 b) was also able to follow the label of $^3$H-galactose which was early concentrated over the Golgi apparatus. There appears thus to be an accumulation of vesicles containing conjugated products found in association with the Golgi apparatus in the axon and at the nerve endings. In describing these vesicles it is difficult to improve on PALAY's (1967) description of synaptic vesicles of which he said that they "like chocolates come in a variety of shapes and sizes and are stuffed with different kinds of fillings."

PALAY (1967) has dealt at some length with different types of vesicles. He notes that the most common synaptic vesicles are about 200–400 Å in diameter and are found in axon terminals at neuromuscular junctions as well as at some other types of nerve endings. Several investigators have found acetylcholine associated with such vesicles. Similarly sized vesicles in the terminals of various adrenergic nerves of the autonomic system have been associated with norepinephrine. Another class of larger vesicles has been shown to contain specific amines and still another type, vasopressin and oxytocin. PALAY notes that certain nerve endings may be characterized by more than one type of morphologically distinctive vesicle and thus may have different stimulatory substances. His analysis of vesicle type and content left the question of why the larger amine-containing vesicles are associated with smaller ones shown by several investigators to contain acetylcholine.

From what has been determined so far (see A. D. SMITH 1971) most neurons are characterized by several types of vesicles including ones that contribute to the lysosomal system. Some are found in the cell body; some in the varicosities and some at the nerve endings. At least the macromolecular components of many of the vesicles appear to undergo part of their assembly

Coordination

Table 2. *Calcium-Dependent Secretion and Exocytosis*

| tissue | approx. diameter of storage vesicle µm | substance secreted | exocytosis electron microscopical evidence | biochemical evidence | references a requirement for calcium b electron microscopic evidence of exocytosis c biochemical evidence of exocytosis |
|---|---|---|---|---|---|
| exocrine pancreas | 1 | digestive enzymes | + | + | a Hokin (1966) b Palade (1959), Ichikawa (1965) c Keller and Cohen (1961) Green, Hirs and Palade (1963) |
| submaxillary gland | 1 | amylase | . | . | a Douglas and Poisner (1963) |
| parotid gland | 1 | amylase | + | . | a Rasmussen and Tenenhouse (1968), Selinger and Naim (1970) b Amsterdam et al. (1969) |
| PMN-leucocyte | 1 | lysosomal enzymes | + | + | a, b, c review by Woodin and Wieneke (1970) |
| adrenal medulla | 1 | lysosomal enzymes | . | + | a, c Schneider (1970) |
| | 0.2 | catecholamines and chromo-granins | + | + | a review by Douglas (1968) b see Grynszpan-Winograd (1971) c review by Kirshner and Kirshner (1971) |
| pancreas (β-cell) | 0.3 | insulin | + | (+) | a Grodsky and Bennett (1966), Milner and Hales (1967) b Williamson et al. (1961), Sato et al. (1966) c Rubenstein et al. (1969) |
| adenohypophysis | 0.25 | LH | + | . | a Samli and Geschwind (1968) b Farquhar (1961) |
| | 0.25 | FSH | . | . | a Jutisz and de la Llosa (1970) |
| | 0.3 | GH | + | . | a MacLeod and Fontham (1970) b De Virgiliis et al. (1968) |
| | 0.6 | MH | + | . | a Parsons (1969), MacLeod and Fontham (1970) b Pasteels (1963), Farquhar (1969) |
| | 0.2 | ACTH | + | . | a Kraicer et al. (1969) b Yamada and Yamashita (1967) |
| | 0.1 | TSH | + | . | a Vale et al. (1967) b Farquhar (1969) |
| | 0.25 | MSH | + | . | a Hopkins (1970 b) b Hopkins (1970 a) |

Table 2 (*continued*)

| tissue | approx. diameter of storage vesicle (μm) | substance secreted | exocytosis | | references<br>*a* requirement for calcium<br>*b* electron microscopic evidence of exocytosis<br>*c* biochemical evidence of exocytosis |
|---|---|---|---|---|---|
| | | | electron microscopical evidence | biochemical evidence | |
| neurohypophysis | 0.15 | oxytocin | + | . | *a* DICKER (1966)<br>*b* NAGASAWA *et al.* (1970) |
| | 0.15 | vasopressin | + | + | *a* DOUGLAS and POISNER (1964), MIKITIN and DOUGLAS (1965)<br>*b* NAGASAWA *et al.* (1970)<br>*c* see UTTENTHAL *et al.* (1971) |
| pericardial organ (crab) | 0.15 | neurosecretory material | . | . | *a* BERLIND and COOKE (1968) |
| platelets (α-granules) | 0.3 | β-glucuronidase | + | + | *a* HOLMSEN and DAY (1968)<br>*b* FRENCH and POOLE (1963)<br>*c* HOLMSEN and DAY (1968) |
| platelets (dense granules) | 0.1 | 5-hydroxy-tryptamine | (+) | (+) | *a* MARKWARDT (1968), TAKAGI *et al.* (1968)<br>*b* WHITE (1968)<br>*c* see HOLMSEN *et al.* (1969) |
| basophil leucocytes | 0.35 | histamine | . | (+) | *a* GREAVES (1969)<br>*c* OSLER (1969) |
| mast cells | 0.4 | histamine | + | + | *a* MONGAR and SCHILD (1958), HÖGBERG and UVNÄS (1960), SURIYAMA and YAMASAKI (1969)<br>*b* HORSFIELD (1965), BLOOM *et al.* (1967)<br>*c* DIAMANT (1967), FILLION *et al.* (1970) |
| cholinergic neurons:<br>skeletal muscle | 0.05 | acetylcholine | (+) | . | *a* review KATZ (1969)<br>*b* HUBBARD and KWANBUNBUMPEN (1968) |
| sympathetic ganglion | 0.05 | acetylcholine | . | . | *a* HARVEY and MACINTOSH (1940) |
| noradrenergic neurons | 0.08 | noradrenaline and chromogranins | + | (+) | *a* SMITH, A. D. (1971)<br>*b* FILLENZ (1971)<br>*c* SMITH, A. D. (1971) |
| | 0.05 | noradrenaline | (+) | . | *a* review by SMITH and WINKLER (1971)<br>*b* FILLENZ (1971) |
| adrenal cortex | — | corticosteroids | | | *a* BIRMINGHAM *et al.* (1953), RUBIN *et al.* (1969) |

Secretion from these tissues depends upon the presence of calcium ions in the extracellular fluid. Evidence consistent with secretion by exocytosis from a storage vesicle is indicated if electron microscopy shows the presence of omega shaped profiles, or if biochemical studies show that other soluble components of the storage vesicle are secreted whereas substances in the cytosol are not released. (For references see A. D. SMITH 1971.)

in the Golgi apparatus in the classical fashion. Perhaps depending upon type
of vesicle, they may be subsequently transported in the axon in membrane-
bounded form. Three aspects of this transport seem to deserve mention:
1. again, substances which might be active internally in the metabolism of the
cell are prevented from doing so because they are separated from the rest
of the cytoplasm by the bounding membrane, 2. their transport seems to be
related as it does in many other instances to the presence of microtubules,
and 3. they may be exocytosed not only at the nerve endings but sometimes
along the course of the neuron.

The complexity of the nervous system and the fact that many of its cells
appear to contain Golgi-derived membrane-bounded vesicles carrying quite
different materials together with the rather wide dispersal of Golgi apparatus
lead to a question that is not answerable in the light of presently available
information: Are there morphologically similar but biochemically quite different Golgi apparatus in nerve cells or is there a tremendous amount of
differentiation which changes in character with development? This question
can be raised with many types of cells, but it is particularly pertinent with
respect to nerve cells because of the long period of development of the system
and its association with progressive changes in capacities.

The presence of lysosomes (HOLTZMAN 1969, 1971) suggests strongly
continuous remodelling and rebuilding processes whether the hydrolases involved are acting intracellularly in the manner of the original concept of
lysosomal function or whether some of them become extracellular and play
a part in the general remodelling of the system.

Whereas the amine content and electron density of the vesicle contents has
allowed a reasonably clear tracing of some types of transmitter substances
from the Golgi apparatus to the region of the synapse, there is a continuing
controversy as to whether the acetylcholine-containing synaptic vesicles are
formed by the Golgi apparatus. Although uncertainty still exists and there
are related questions of the possible recycling of membrane-bounded components within the synapse (for example see HOLTZMAN et al. 1971, DOUGLAS
et al. 1971, HEUSER and REESE 1973, CECCARELLI et al. 1973) some evidence
(see GRAY 1970, BRUNNGRABER 1972 a, GRIFFITH and BONDAREFF 1973) does
suggest origin from the apparatus. This is in keeping with the general
acceptance of the role of the Golgi apparatus in various secretory processes
and the substantial agreement by those authors whose papers were published
in the *Philosophical Transactions of the Royal Society of London* (1971)
that release of transmitter substances largely involves exocytosis. It is of some
interest to note that work leading in the direction of transmission by exocytosis was done by DEL CASTILLO and KATZ in the early 1950's
(see DEL CASTILLO and KATZ 1954, 1956). KATZ summarized the
work in his Croonian Lecture in 1962. The lecture was primarily electro-
physiological, but it held that materials were released in quantal fashion
rather than by osmotic phenomena.

Quantal transfer of transmitter substances could explain many aspects of
the functioning of the nervous system including the rapidity with which
stimuli may be passed through segments of the system. It does not invalidate

some of the earlier physiological characterizations of the membrane though it does bring them somewhat more into line with the behaviour of membranes of other cells.

One of the very general characteristics of cellular secretion is its dependence upon the presence of calcium in the extracellular fluid. This dependence obviously relates to physiological characteristics of the membrane. A. D. SMITH (1971) has presented a summary table of the extensiveness of this phenomena and included the neuronal membranes (Table 2).

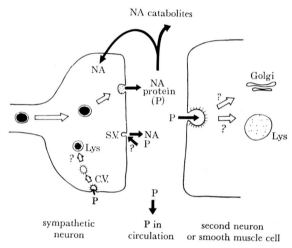

Fig. 80. Diagram of protein secretion from a sympathetic neuron. For explanation see text and SMITH (1971). From A. D. SMITH, Phil. Trans. Roy. Soc. (Lond.) B **261**, 1971.

In summary, it is of interest to compare SMITH's interpretation of the movement of proteins in neurons and the involvement in the process of both exocytosis and endocytosis. In its assumptions of the activity of specific organelles it closely resembles comparable activities in other sorts of cells in which intercellular materials become important in the determination of cellular functioning (Fig. 80).

## B. The Catecholamines

Much of the information about the functioning of neurons is based on the study of adrenergic neurons. This in turn has been enhanced by the investigation of cells of the adrenal medulla, which also synthesize, store, and secrete catecholamines. The comparability of these cells to certain neurons has been firmly established (for references see KIRSHNER and KIRSHNER 1971). The biological processes involved in granule formation appear to be identical as do the release processes, notably those pertaining to the simultaneous release of catecholamines and adenine nucleotides, chromogranins, and dopamine β-hydroxylase. The large chromaffin granules of the adrenal medulla have been isolated and analyzed chemically. There seems to be the same transfer of the protein components, the same involvement of an assembly process in the Golgi apparatus, and finally secretion by exocytosis.

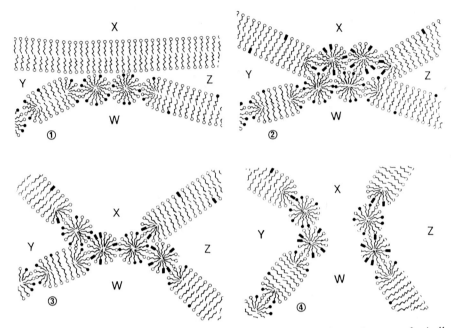

Fig. 81. Diagram showing a theoretical interpretation of the possible involvement of micellar configurations in membrane fusions that could be involved in exocytosis or endocytosis. The letters indicate regions of the intra- and extracellular space definable specifically only in terms of the sort of membrane fusion involved. They do, however, indicate changes in interrelationships with fusion and separation. The numbers suggest progressive interactions between and changes in organization of both the bimolecular leaflet and micellar segments of membranes. WINKLER (1971) has interpreted this as a possible illustration of the involvement of lysolecithin in the fusion process, and LUCY (1969) has noted that lysolecithin is a surface-active molecule. From LUCY, in: Lysosomes in biology and pathology, Vol. 2 (DINGLE, J. T., and H. B. FELL, eds.). Amsterdam: North-Holland. 1969.

Exocytosis had been suggested some time ago by both morphological and biochemical studies. SCHNEIDER *et al.* (1967) reviewed this evidence and presented further biochemical data consistent with this idea. They also noted the importance of $Ca^{++}$ in the process and suggested that this is the only mechanism that would allow the simultaneous release of both the low and high molecular weight components of the chromaffin granules. Exocytosis would also involve interactions between the membrane of the granule and the plasma membrane, and WINKLER (1971) concluded that modifications in these membranes might be necessary for fusion (Fig. 81). Attention was given to the modifications of both the vesicle membrane and the plasma membrane, and WINKLER (1971) suggested that changes in membrane structure might have to do with fusions between the membranes. Comparable pathways appear to hold for the adrenergic neurons. A notable feature of some considerable consequence in medicine has been the extent to which drugs capable of reducing the production of, or modifying the activities of, the catecholamines have been developed.

WINKLER (1971) has considered chromaffin granules in relation to a cyclic

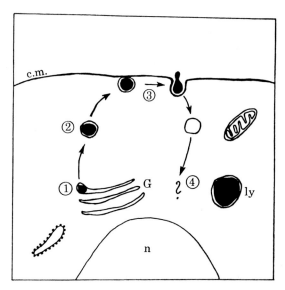

Fig. 82. Diagram of possible behavior of chromaffin granules, including exocytosis followed by endocytosis. $G$ = Golgi apparatus, $ly$ = lysosome, $n$ = nucleus, $c.m.$ = cell membrane. From WINKLER, Phil. Trans. Roy. Soc. (Lond.) B **261**, 1971.

process (Fig. 82) in which they originate from the Golgi apparatus, are exocytosed, and their membranes are then returned to the cytoplasm. He has noted that freeze-etch preparations of this material demonstrate minute granules and plaques not ordinarily seen in materials otherwise prepared for electron microscopy. In the other studies this has turned out to be true of additional membranes in the cell giving to these structures a more complex character as organelles composed of numerous specific molecular determinants than is suggested by preparation using standard fixatives which tend to create a uniformity of appearance in membranes.

The soluble contents of the granules: catecholamines, ATP, and chromogranins can be removed and the membranes of chromaffin granules recovered. When this is done the composition of the membranes is found to be as in Table 3:

Table 3. *Composition of Membranes of Bovine Chromaffin Granules*

| | |
|---|---|
| protein | 1 mg |
| lipid-phosphorus | 2.4 μmol |
| cholesterol | 1.66 μmol |
| $Mg^{2+}$-ATPase | 1.8 μmol/h |
| dopamine β-hydrolase | 0.012 μmol/h |
| (substrate 5 μmol/l) | |
| cytochrome b-599 | 0.46 ($E_{425}$) |
| chromogranin A | 0.04 mg |

[The data are taken from WINKLER et al. (1970). $Mg^{2+}$-ATPase was measured at the optimal pH of 6.4; dopamine β-hydroxylase was determined at pH 5.5. Chromogranin A was assayed by microcomplement fixation.]

One importance of the work on the chromaffin granules is that immunochemical methods have been applied specifically to chromogranin A and
dopamine β-hydroxylase to identify their location in the cell. Studies of
immunofluorescence have indicated that these two proteins correspond to the
distribution of catecholamine vesicles. Further extension of these techniques

Table 4. *Lysolecithin Content of Chromaffin Granules from the Adrenal Medulla
of Various Species*

| adrenal gland | lysolecithin (% of total lipid-P) | references |
|---|---|---|
| ox | 16.8 | BLASCHKO *et al.* (1967 a) WINKLER *et al.* (1967 b) |
| ox | 12.9 | TRIFARÓ (1969) |
| horse | 7.1 | WINKLER *et al.* (1967 b) |
| rat | 15.4 | WINKLER *et al.* (1967 b) |
| pig (total chr. gran.) | 11.3 | WINKLER *et al.* (1967 b) |
| pig (mainly nonadrenaline-containing) | 19.7 | WINKLER (1969) |
| human phaeochromocytoma (3) | 11.7, 17.8, 23.8 | BLASCHKO *et al.* (1968) |

(The references included in this table are not in the bibliography; see WINKLER 1971.)

Table 5. *Amino Acid Composition of Insoluble Proteins from Bovine Chromaffin Granules
(in percent by mass)*

| | | | | | |
|---|---|---|---|---|---|
| Lys | 6.1 | 8.0 | Gly | 4.2 | 5.0 |
| His | 2.8 | 2.7 | Ala | 5.3 | 5.8 |
| NH3 | 2.0 | 4.2 | Cys | 0.5 | 1.4 |
| Arg | 7.1 | 8.1 | Val | 5.0 | 6.3 |
| Asp | 9.6 | 8.5 | Met | 2.9 | 1.6 |
| Thr | 4.6 | 3.3 | Ile | 3.7 | 3.9 |
| Ser | 7.6 | 5.0 | Leu | 9.9 | 9.9 |
| Glu | 15.4 | 13.9 | Tyr | 4.1 | 3.5 |
| Pro | 5.4 | 4.4 | Phe | 5.5 | 4.1 |

[The data from the first columns of numbers are taken from WINKLER *et al.* (1970); those
from the second columns from HELLE and SERCK-HANSSEN (1969); see WINKLER (1971).]

should allow the site-to-site tracing of precursors in the assembly and hope-
fully the relation of assembled macromolecules to the cell surface (LIVETT
*et al.* 1971).

The significance of work on the direct analysis of membranes is to illustrate
that some of the active principles involve a very intimate relationship between the molecular composition of membranes derived from the Golgi
apparatus and materials that are assembled and synthesized as components
of classic secretory products. The dopamine β-hydroxylase appears to be
about equally divided between the soluble fraction and the membranes.
WINKLER considers the possible function of the various membrane components

although he supposes that they have largely to do with the uptake of catechol-amines and their discharge. Lysolecithin occurs in the membranes of chromaf-fin granules in unusually high amounts (Table 4), whereas it occurs in only small amounts in the membranes of mitochondria and microsomes.

A possible importance of the high lysolecithin content lies in LUCY's (1969) postulate that lysolecithin may play a particular part in the fusion processes of granule membranes and the cell membrane. To these analyses WINKLER has added one dealing with the amino acid composition of insoluble proteins in chromaffin granules (Table 5). Such an analysis is important because it anticipates later work giving particular compositional character to the mem-branes of secretion products derived from the Golgi apparatus and because the difference between this character and the earlier stages in the ontogeny of membrane development lends credence to SJÖSTRAND's (see 1968) concept that aspects of membrane specialization are in a sense a primary function of the Golgi apparatus.

There still needs to be in some systems a comparison of the compositional and structural characteristics of membranes responsible for fusion with the cell membrane in the transport of extracellular materials and what must be a comparable specialization of membranes that are destined to fuse with elements of the lysosomal system in the transfer of hydrolases. The impor-tance of the production of catecholamines by cells of the adrenal medulla is twofold. It provides a large supply of active substances by which the functions of the organism can be correlated, a supply large enough to permit coping with many situations of stress, and it provides a ready example of the manner in which comparable activities are conducted by somewhat more difficult-to-study cells of the nervous system. For some additional studies of the cellular components involved in the synthesis and secretion of biogenic amines see EKHOLM and ERICSON (1968), RÖHLICH et al. (1971), RUBIN et al. (1971), COGGESHALL (1972 a), ERICSON (1972), and GERSHON and NUNEZ (1973).

# IX. Fertilization

There have been numerous studies that indicate that the structure recog-nized as the Golgi apparatus plays certain roles in the fertilization of organisms that range from single-celled to animals as complex as mammals. Some of the factors involved are discussed in other sections of this work. Although information is still very incomplete, reference must be made to the involvement of the Golgi apparatus in the formation of the acrosome and some of the structures protecting the ovum (for general information see BACCETTI 1970, BIGGERS and SCHUETZ 1972).

## A. The Acrosome

The acrosome is a cellular component developed during spermiogenesis. A few details of this involvement are now becoming clear, as is the fact that the Golgi apparatus is a very active and sometimes much modified structure during oogenesis. The studies of syngamy which go beyond purely

morphological considerations suggest that differentiated activities of the
organelle may be the basis for numerous steps in fertilization, and the studies
that have been done on organisms at different levels on the evolutionary scale
indicate that such properties as the Golgi apparatus imparts to cell surfaces
may be of fundamental importance in determining genetic compatibility.
As suggested elsewhere, participation in formation of the acrosome was one of
the earliest studied manifestations of Golgi apparatus activity, although for
a long time the association was difficult to define. Sperm are characteristically
very small and their components could not readily be studied by the
procedures of the early cytologists. As a consequence, the mechanics of
participation of the Golgi apparatus in acrosome formation did not receive
a fraction of the attention given to, say, its role in secretion.

Nonetheless, BOWEN began his extensive studies of the Golgi apparatus with
a series of investigations of its involvement in acrosome formation. Though
he later became concerned primarily with its function in processes of glandular secretion, he confirmed the variation in its patterns of activity in acrosome
formation in different organisms. WILSON (1925) was to comment that
although much studied by many investigators, the acrosome was treated with
"rather scanty respect", but that BOWEN (1924) suggested the possibility that
the acrosome might contain some substance(s) which played a part in the
activation of the fertilization process. The other suggestion inherent in the
early studies is that acrosome development involves differences that are
correlated with evolutionary level (for more recent information, see YASUZUMI
1974).

Examples of more modern studies of acrosome formation at quite different
levels in the evolutionary scale are that of KAYE (1962) on acrosome formation in the house cricket (see also Fig. 37) and that of SANDOZ (1970 a) and of
SUSI et al. (1971) on successive changes in the Golgi apparatus during development of the spermatid in the mouse and the rat. The general modern literature
has been summarized by DAN (1970). Without going into the details of
differences among species, one is concerned, as with other Golgi apparatus
products, with both the membrane and the contents. In the latter case,
BOWEN has been proven correct in the assumption that some analogy exists
between acrosomal content and that of vesicles produced by the Golgi
apparatus in secretory cells.

The acrosomal membrane, which comes ultimately to bound the anterior
end of the sperm during sperm activation, has been diagrammed by SUSI
et al. (1971) as being derived from membranes of vesicles from the Golgi
apparatus (Fig. 83). Because of its obvious importance in fertilization, this
membrane has been extensively studied. COLWIN and COLWIN (1967) have
shown a considerable number of the variations in this membrane and its
activities during the fertilization process. A number of authors (see DAN
1970) subscribe to the idea that there are significant differences between the
original plasma membrane of the sperm and specialized portions of it making
up the so-called process membrane, derived in part at least from the acrosome.
With some cytochemical stains there do not appear to be discernible differences
in the two portions of the membrane. On the other hand, it has been

established by EDELMAN and MILLETTE (1971) that in the mouse the plasma membrane overlying the acrosomal region contains large numbers of concanavalin A binding sites per sperm. Though clear evidence is not available, it seems reasonable from all the other data to suggest a role for the Golgi apparatus in the organization of such a membrane, thickened by a coat or otherwise, and containing a large number of carbohydrate moieties which may function in recognition and binding processes. METZ (1967) has interpreted

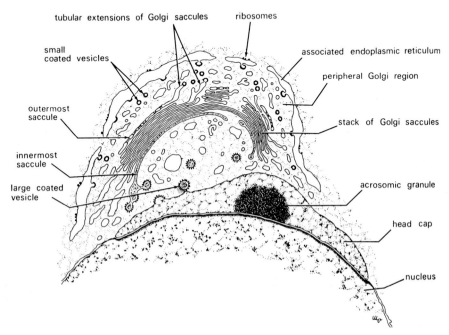

Fig. 83. Diagram showing various Golgi apparatus components and the forming acrosomal granule in an early-stage spermatid in the rat. The uppermost Golgi elements are the proximal face. The vesicles contributing to the acrosomal membrane are mostly from the distal face. From SUSI et al., Amer. J. Anat. **130**, 1971.

the recognition and binding between egg and sperm as due to a series of properties, including one that has not been extensively elucidated—the presence of antigens, or substances with antigenic properties, on the membrane surface. W. A. ANDERSON (1968) also suggested that the cell surfaces might have characteristics of importance in an immunological sense (for more recent evidence see KOEHLER and PERKINS 1974). Either or both interpretations might find support in the concept that the acrosomal membrane and the portions differentiated from it are transferred from a functional Golgi apparatus. The question of whether or not segments of the sperm membrane enter the ovum, and if so, what functions they may mediate, is still a matter for exploration (DAN 1970).

Other questions about acrosome activity in fertilization are related to the contents. LOEB (1901) appears to have been first to suggest that some lytic substance or substances were carried by the sperm, and this suggestion implied that the acrosome might be a carrier of substances involved in breaking down the protective structures at the egg surface.

For a long time principal concern for substances of lytic character in the acrosome was concentrated on hyaluronidase (see DOTT 1969). Hyaluronidase had long been recognized as a polysaccharase, and the possibility that it operated on portions of the structures surrounding the ovum gave an aspect of chemical activity to what was once thought to be a purely mechanical entry process. Then, in 1963, DE DUVE postulated that lysosomal enzymes generally might play a part in the reaction involving the acrosomal contents and the material surrounding the egg. Subsequently, evidence has accumulated to indicate that in one sense the acrosome can be interpreted as a "giant" lysosome, the active lysins of which play a part in the complex processes involved in the breakdown of the egg's protective structures (see McRORIE and WILLIAMS 1974). The failure of some investigators to associate this activity with a specific secretory function of the sperm loses much of its meaning with the accumulating evidence that lysosomal enzymes are frequently released to function outside the cell membrane.

It seems unlikely, however, that such activity represents the only function of the Golgi apparatus in fertilization. A number of investigators (see DOTT 1969) have demonstrated in reproductive cells the presence of the same transferases concerned with the assembly of glycoproteins in the Golgi apparatus of other types of cells, and SUSI et al. (1971), following up earlier work by LEBLOND and his colleagues, have made a detailed analysis of changes in the Golgi apparatus observed during maturation of rat spermatids. They have demonstrated patterns of protein transfer from the endoplasmic reticulum and the formation of glycoproteins in the Golgi apparatus quite comparable to those which take place in other types of cells. These observations give the acrosome, long thought to be primarily glycolipid, a high content of glycoprotein. Some of these conjugated compounds have lytic properties and play a part in the breakdown processes of protective substances and membranes, as well as possibly playing a part in the recognition processes which represent the initial step in fertilization.

Early investigators, including BOWEN (1924), were convinced that once the Golgi apparatus had played its part in forming the proacrosome, the organelle had no other function and was subsequently shed from the developing sperm. Electron microscopy has revealed that this is not always the case, and that it may remain to perform some further functions (SANDOZ 1970 b), just as certain other organelles remain in the fully developed sperm though they may not enter the ovum at syngamy. The Golgi apparatus may also produce other structures which play a part in fertilization, as for example in spermiogenesis in the tick *Dermacentor andersoni* (REGER 1974).

There are a few organisms without acrosomes (COLWIN and COLWIN 1967). Among them are forms which have no conspicuous protective structures surrounding their gametes and ciliated forms. It seems reasonable to assume that recognition in such species is a characteristic of particular features of the surface. Observations relating to this are dealt with in Section XIII. Naked gametes obviously carry no requirement for lytic breakdown of protective structures.

It would seem that the acrosome represents another instance of membrane packaging a highly active product or complex of products. Whether an acrosome is developed in a particular species appears to be correlated with the activity of the Golgi apparatus in developing protective structures that come to surround the egg. If such structures are developed, then the acrosome appears to function like other Golgi products in both recognition and activities that modify cellular associations.

## B. The Egg Golgi Apparatus

The eggs of mammals are surrounded by a zona pellucida of varying thickness (see ZAMBONI 1971), while those of other species are surrounded by various membranes which may constitute a counterpart of the zona pellucida or, in a few instances, are without such protective layers. There is some confusion of terminology, since "vitelline membrane" has been applied to the plasma membrane in mammalian eggs, and such expressions as "vitelline coat" and "vitelline envelope" have been applied to other structures in other species (see COLWIN and COLWIN 1967). Here it will suffice to point out that these external structures, which may sometimes show a considerable degree of complexity, appear in many cases to be formed by activities carried out in part in the Golgi apparatus. Fig. 84 shows the zona pellucida around a normal mammalian oocyte. Although the origin of this coating has not been clearly elucidated, it seems likely that activities of the Golgi apparatus either of the follicle cells or of the oocyte are involved. At one stage in the formation of the mammalian oocyte, the Golgi apparatus is a typical-appearing structure consistently found in a juxtanuclear position. Later it becomes very much modified (Fig. 85), and it is from this modified state that materials are released (BACA and ZAMBONI 1967). Extensive modifications have also been reported in other species.

Fig. 86 illustrates the endochorion and the exochorion of the dragonfly egg. In this case there is clear evidence that these protective structures are secreted by the oocyte Golgi apparatus at temporally distinct stages in oocyte maturation. BEAMS and KESSEL (1969) have noted that this represents a change in the activity of the apparatus at some specific stage in development. They consider this unique in that at that time changes in activity had been emphasized only by BAINTON and FARQUHAR (1966) in developing leucocytes. Actually, developmental changes in the activity of the Golgi apparatus are, as is pointed out elsewhere in this work, common enough to be considered one of its characteristics. The BEAMS and KESSEL study of the dragonfly egg surface is, however, of particular interest in that the substance of some of the secretion bodies appears to have a distinctive morphology, which is altered once the material gets outside the plasma membrane.

Annulate lamellae have been studied in many types of cells, both normal and abnormal (see KESSEL 1968). They have, however, received special

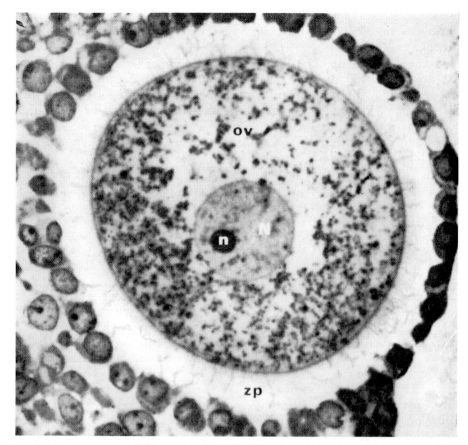

Fig. 84. Human oocyte (*ov*) in an antral follicle. The oocyte is surrounded by corona radiata cells (not marked), and within their circle by a zona pellucida (*zp*). $N$ = nucleus, $n$ = nucleolus. From BACA and ZAMBONI, J. Ultrastruct. Res. **19**, 1967. ×940.

attention in the development of oocytes. A tentative suggestion based on the similarity of membrane arrangement and vesicle production is that annulate lamellae might represent a supplement to the Golgi apparatus. KESSEL (1968) quotes HSU (1963) as indicating that in a tunicate the annulate lamellae are developed by a process involving both nuclear membranes, and he summarizes their development and cites the work of many investigators to support annulate lamellar development from the outer nuclear envelope, though he notes that a different pattern takes place when they are intranuclear in character. A relationship between the Golgi apparatus and structures derived from the nuclear envelope finds some parallel in the instances where the organelle is itself closely related to the nuclear envelope. For example, KESSEL (1971) following developmental stages in the grasshopper embryo has suggested that the Golgi apparatus is derived from the nuclear envelope. Whether there actually is a significant relationship in origin and/or function between the annulate lamellae and the Golgi apparatus is not clear. KESSEL

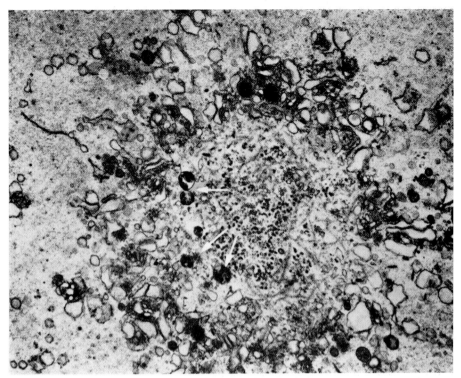

Fig. 85. Modification of the Golgi apparatus in a maturing human oocyte. The structure appears to be producing cortical granules. From BACA and ZAMBONI, J. Ultrastruct. Res. **19**, 1967. ×6,400.

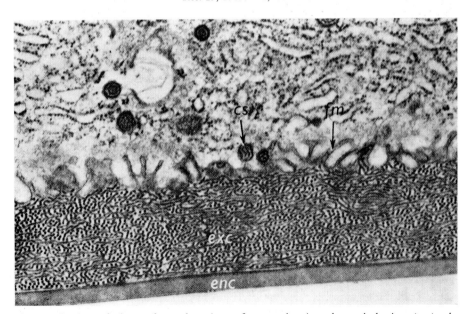

Fig. 86. Section of the surface of a dragonfly egg showing the endochorion (*enc*), the exochorion (*exc*), secretion bodies (*cs*), and membrane folding (*fm*). From BEAMS and KESSEL, J. Cell Sci. **4**, 1969. ×31,500.

is certainly correct in his statement that the Golgi apparatus is being increasingly recognized for functioning in harmony with other cytomembrane systems.

## C. Syngamy

It was once thought that the sperm literally bored its way into the egg cell (see DAN 1967). Mechanical activity on the part of the sperm still appears a possibility, but only with considerable associated chemical activity. The calcium dependence of fertilization (DAN 1967) is in accord with the calcium dependency of other Golgi apparatus activities (see Table 2). The viability and motility of the sperm differ considerably even within a single evolutionary group, as does the initiation of contact between sperm and egg. To take a single example, W. A. ANDERSON (1968) has shown that in the sea urchin an early development is the appearance of very fine fibrils between the acrosomal coat and the egg surface, but whether this is a factor in recognition or in the initiation of contact processes is not clear. A more generally emphasized reaction (COLWIN and COLWIN 1967) is the development of an acrosomal process which contacts the egg cell at some specific point. Originally thought to be strictly an acrosomal reaction, there is apparently a mutual reaction between the egg and sperm breaking down the egg surface and the acrosome. This reaction appears to involve a fusion of the acrosomal and egg plasma membranes. The sperm nucleus enters the egg and initiates formation of the zygote.

Though there was a misinterpretation involved, it was GOLGI's eighteenth century predecessor SPALLANZANI who demonstrated the spermatozoon. It was late into the period of the study of the acrosome before it became clear that a product of the Golgi apparatus was responsible for the introduction of the sperm nucleus into the egg to effect fertilization, and that this was basically a chemical process and not merely a mechanical process of perforation and penetration. If there are any doubts about the influence of the ancient University of Pavia, they can be dispelled by the fact that another of its greats, VALLISNERI, had a part in explaining what is now curiously set forth (TYLER 1967) as the "philosophy of fertilization". What seems most important for this work is that certain reactions of the Golgi apparatus are responsible for forming the coating that protects the egg from fertilization by any except a selected few sperm. Other reactions of the Golgi apparatus are responsible for producing those processes that bring about the breakdown of this protective coating and facilitate the entry of compatible sperm. It would seem that in this process the Golgi apparatus has, but in a highly specialized fashion, the same sort of determinative control over the association and development of the early stages of organisms that it has over their later stages. It seems quite capable of acting in egg or sperm in a manner that will translate the genome and its modifiers into effective forces in the determination of cellular events by the characterization of surface materials.

It has taken a long time to reassociate the Golgi apparatus with one of its functions that was apparent very early, and an even longer time to discover the details of its involvement in the maturation of the oocyte. Given the information now available, however, there seems to be no reason for assuming that it doesn't play the same roles in membrane and associated material production and in the contribution of lysosomal enzymes in reproductive phases that it plays in the somatic tissues of the organism. It contributes both determinants and active substances, and if by chance there is at fertilization a transfer as well as a fusion of membranes, it may well contribute factors modifying certain stages of activity of the genome. Its function in fertilization and the possible interrelationship of it to other membrane systems reemphasizes its role as a membrane specialization center, which is a significant factor in determining the cell associations.

## X. Other Functions

### A. Lysosomes

Of the cellular organelles, the lysosomes entered the scene late and indirectly. DE DUVE (1969), who made the initial discovery, dates their introduction on December 16, 1949. He and a few colleagues, working in what he describes as a derelict laboratory on another problem, found them more or less accidentally in the form of latent acid phosphatase activity.

DE DUVE said he had a hunch that the latency characteristics of his unexpected Christmas present for 1949 would prove of importance if they could work out the facts. Actually it took only a few months to relate the latency to a membranelike barrier between the enzyme and its substrate. As he tells the tale of the isolation and characterization of the lysosome in retrospect he makes much of his hunches and mistakes that ultimately worked to the benefit of a careful definition. While DE DUVE and his colleagues were conducting their biochemical studies, NOVIKOFF (see DE DUVE 1969) was studying cellular entities by morphological and cytochemical techniques, including staining procedures for acid phosphatase.

It turned out later from the work of a number of investigators that there are several hydrolytic enzymes involved. Some of them seem to complement each other in activity and for a while lysosomes were defined as lytic particles containing these specific enzymes. Their definition, however, continued to be based largely upon biochemical properties until finally NOVIKOFF managed to take some electron micrographs of particles in which he could demonstrate acid phosphatase cytochemically and which were membrane-bounded, verifying DE DUVE's interpretation of the latency phenomenon. Defining the role of the lysosomes presented some more difficulties and a definition did not come until more was understood about their formation and certain activities of the cell.

Part of the definition of these came from the studies of COHN and FEDORKO (1969) who worked on heterophil granulocytes and macrophages, both of which are distinguished by much endocytic activity and contain many lysosomes. It could be demonstrated with markers such as peroxidase that in such cells, pinocytic vacuoles are common but without acid hydrolase activity. On the other hand, certain cisternae on the distal face of the Golgi apparatus do contain such enzymes and smooth endoplasmic reticulum which comes in the vicinity of the distal face (GERL) may also contain them. As noted, the designation, GERL, intended to emphasize a relationship of the Golgi apparatus, the endoplasmic reticulum, and the lysosomes, was first set up by NOVIKOFF and his co-workers (see HOLTZMAN et al. 1967). GERL is now generally interpreted by NOVIKOFF as indicating a modification of the enzymatic content of profiles of smooth endoplasmic reticulum coming into the vicinity of the Golgi apparatus (NOVIKOFF et al. 1971). The observations of COHN and FEDORKO (1969) were consistent with the idea that vesicles derived from the Golgi apparatus or the smooth endoplasmic reticulum may constitute primary lysosomes; that is, may contain essential hydrolytic enzymes.

COHN et al. (1966 a, b) made a relatively early investigation of the transfer of smooth vesicles to the neighborhood of pinocytic vacuoles and later explored their ultimate fusion. They had presented the following figure (Fig. 87) as a scheme for the formation of primary and secondary lysosomes in the macrophage. DE DUVE had earlier also listed phagocytosis as one of the possibilities for the origin of the granules. He went on to concern himself with their functioning. In this relationship he went back to some 19th century work by METCHNIKOFF (for references see DE DUVE 1969) who in 1865 had described the phenomenon that came to be known as autolytic digestion. There arose from these deliberations the concept that the functions of the lysosome might include autolytic degradation in which components of the cell, no longer serving a useful purpose, were broken down into their building blocks for return to the metabolic activities of the cell. For a long time this was thought to be the principal function of the lysosomes in animal cells other than the obviously phagocytic unicellular forms.

With the growing indications that endocytosis (phagocytosis or pinocytosis) is a general function common to virtually all animal cells, primary attention turned toward the lysosomal degradation of materials brought in from outside of the cell. Functional lysosomes are formed by the fusion of invaginated membranes sometimes surrounding ingested materials with membrane-bounded vesicles containing hydrolytic enzymes derived from the Golgi apparatus (or GERL). The molecular products of digestion are largely transported to the cytoplasm.

In some types of cells there are membrane-bounded bodies which contain smaller membrane-bounded bodies referred to as multivesicular bodies (MVB) (Fig. 88). These have been investigated in detail by FRIEND (1969) in rat epididymal tissue. He used cytochemical stains for acid phosphatase and thiamine pyrophosphatase to indicate derivatives from certain Golgi cisternae and the acid phosphatase of the lysosomal contents and staining of horseradish peroxidase as a marker to indicate cell surface-derived material. His conclusion was that the larger outer membrane was frequently derived endocytically and the smaller inner vesicles derived from the Golgi apparatus. On the basis of these cytochemical tests he did not arrive at a specific definition of the nature of these bodies but supposed they might represent a specialized component of the lysosomal system. The original concept of the lysosome as essential to the maintenance of certain balances within the cell seems still valid since changing conditions within the cell bring about changing balances among its components. There is no doubt about the activity of components in the lysosomal system in such cases as the azurophil granules

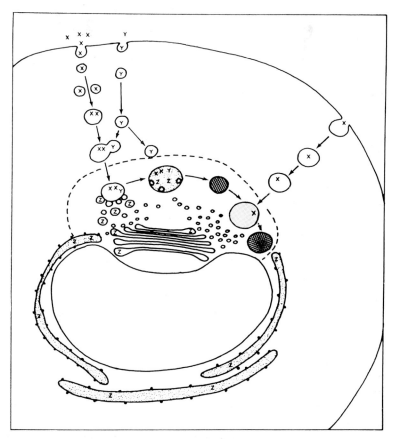

Fig. 87. Diagram of the formation of primary and secondary lysosomes showing molecules X and Y taken up by pinocytosis and enclosed within segments of the plasma membrane. These may fuse and migrate to the Golgi zone. There is then a transfer of hydrolases (Z) to these vesicles from the Golgi apparatus to form functional lysosomes. From COHN and FEDORKO, in: Lysosomes in biology and pathology, Vol. 1 (DINGLE, J. T., and H. B. FELL, eds.). Amsterdam: North-Holland. 1969.

of leucocytes where foreign materials are introduced into the cell. But the lysosomes are important in other senses, too. As an example it is worth citing again their role in the involution of the mammary gland cell (WOESSNER 1969). For extensive information on the lysosomal system and the characteristics of lysosomal functions in various cell types see DINGLE and FELL (1969) and DINGLE (1973).

More recently it has become clear that at least some lysosomal enzymes may make their way out of the cell (DINGLE 1969, HELMINEN and ERICSSON

1970 b, Wright and Malawista 1972) and could play a significant part in the metabolic activities that modify development of intercellular materials. This is becoming a large area of investigation, but inasmuch as the focus here is on the Golgi apparatus and its participation in the formation of the lysosomes it will not be treated in any detail, though it is of obvious importance in any consideration of general cellular functioning.

The lysosomes in relation to disease present an important concept for it directly involves the Golgi apparatus and its functioning in a wide variety of disease conditions. Though the details of the participation of various cellular components are not known, the early 1960's saw the lysosomes associated with a series of so-called lysosomal diseases (de Duve 1969). de Duve discusses a number of so-called congenital storage diseases in which children die at an early age with unusual accumulations of storage products. Certain of these were found to be related to missing enzymes, the absence of which constituted lysosomal defects. Later, it was discovered that various types of injury and drug treatment modified lysosomal function and caused serious metabolic disturbances including various degenerative diseases.

Since these early observations, lysosomes have been implicated in a number of other congenital diseases (Hers and van Hoof 1969). Increase in the number of lysosomes and effects of lysosomes on infecting agents and drug treatments have become of increasing importance, and direct effects of lysosomal hydrolases have been called responsible for many fundamental modifications of normal cellular activity. Even a brief exploration of these functions is beyond the scope of this review except for an emphasis on the production of their active substances by the Golgi apparatus. Since, however, comparable enzymes have been demonstrated in plant cells not known to be involved in endocytosis or any disease processes it may be necessary to consider additional functional roles of the lysosomal system.

From the outset the existence of lysosomes in plant cells has presented unanswered questions. Poux (1963, 1965, 1970), Dauwalder et al. (1969), Coulomb and Coulon (1971), and others have demonstrated the existence of acid phosphatase in certain plant cells, generally finding along with Matile (1969) a distribution of the enzyme in vacuoles as well as in the Golgi apparatus in some cell types. Matile has interpreted the plant vacuole as having lysosomal activity although some other investigators (see Matile 1969) have indicated more or less specific cellular components as being characterized by this activity. Poux (1965), Matile (1969), and others have also shown acid phosphatase by cytochemical methods in connection with the aleurone grains and other storage products of certain types of plant cells.

Dauwalder et al. (1969) have studied this problem from the standpoint of the Golgi apparatus in different cell lineages of the corn root tip showing that in certain types of cells there appears to be acid phosphatase developed toward the distal face of the Golgi stack. There is even in some plant cells (Fig. 89) what seems to equate to Novikoff's GERL with a distribution of acid phosphatase activity in what appear to be segments of smooth endoplasmic reticulum in the vicinity of the Golgi apparatus.

At this time one can only conclude that there is lysosomal activity in plant

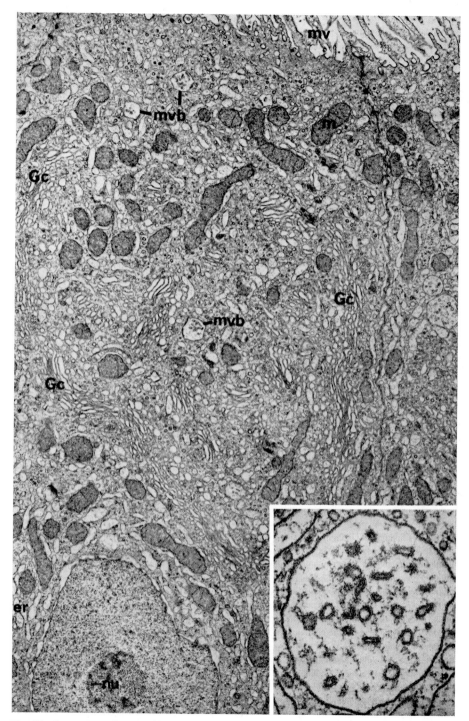

Fig. 88. A portion of rat epididymal tissue. The inset shows a large multivesicular body. For explanation see the text. *er* = endoplasmic reticulum, *Gc* = Golgi apparatus, *m* = mitochondrion, *mvb* = multivesicular body, *nu* = nucleolus, *mv* = microvilli. From FRIEND, J. Cell Biol. **41**, 1969. Courtesy of Rockefeller University Press. ×13,200, inset ×77,500.

cells but that the form in which it occurs is not cytologically directly comparable to that in animal cells.

In the continuing search for the fundamental role of the Golgi apparatus in cellular metabolism an early conflict was seen with the clear relationship of the apparatus with secretion and its general occurrence in cells including those not known to be differentiated for secretion. As an extension of DE ROBERTIS' (1964) suggestion that all neurons are secretory it may be true that all cells are in some sense secretory. It has been suggested (see WHALEY et al. 1972, DAUWALDER et al. 1972) that much of the plasma membrane

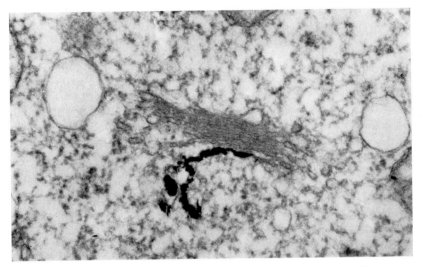

Fig. 89. A cytochemical demonstration of acid phosphatase in a plant cell, GERL.
From M. DAUWALDER, Cell Research Institute, University of Texas at Austin. ×60,000

may be derived from the Golgi apparatus and that the apparatus may play a critical part in the synthesis of carbohydrate components of either the membrane or its associated *coat* materials (see also BENNETT et al. 1974). The turnover of surface materials is probably of general occurrence and could provide a common role for the apparatus in most, if not all, cell types. The reinterpretation of the secretory functions of the Golgi apparatus to provide even this very broad base is, however, perhaps not sufficient to explain the general occurrence of the organelle and some of its other functions. Among these is its involvement in the production of hydrolytic enzymes. As an example the production of acid phosphatase finds a neat explanation of its function in the activity of the lysosomes in many types of animal cells. The lysosomal system may have a rather broad role in the continuing development and changing activities of cells and associated surface components allowing the cells to create and change cell-to-environment and cell-to-cell relations. But the production of hydrolytic enzymes is also a conspicuous feature in certain types of plant cells where the lysosomal system, if actually present, may not have the definitive characteristics of the animal cell system. Currently, further speculation seems unwise, but it does seem clear that the organelle plays important parts in both anabolic and catabolic phases of cellular metabolism.

## B. Leucocytes

Leucocytes arise in the bone marrow from cells generally known as myeloblasts. They have been a favorite object of study for a long time and as is usual in such cases different terminologies have been applied to their successive stages of development. In discussing them, it seems most appropriate to use the terminology of particular investigators but to introduce some comparable terms where essential. Myeloblasts are characterized by the usual organelles but notably by sparse endoplasmic reticulum and relatively small Golgi apparatus. There do not appear to be morphologically different types of myeloblasts, but the population of myeloblasts differentiates into various types of granulated leucocytes as well as into some other types of cells.

Granulated leucocytes include heterophils or neutrophils, eosinophils, and basophils, distinguishable on the basis of staining reactions and granule content.

BAINTON and FARQUHAR (1966) have worked out the developmental stages of heterophils though they have also shown some characteristics of eosinophils and basophils (Fig. 90). They found that as a myeloblast becomes a progranulocyte there is a notable development of the Golgi apparatus which produces granules classified by their staining reaction as azurophil granules. With time the progranulocyte ceases to produce azurophil granules and differentiates into a myelocyte. With this differentiation the activity of the Golgi apparatus changes and it produces less dense, so-called specific granules. Since azurophil granules are produced only for a limited period of time, cell divisions, which continue during differentiation, diminish the number of the azurophil granules per cell so that mature leucocytes contain more specific granules than azurophil granules. SUTTON and WEISS (1966) using the chicken; and later NICHOLS et al. (1971) using the rabbit, the guinea pig, and human beings; and other investigators have all emphasized the reduction of the number of azurophils associated with phagocytosis and the failure of the cells later to produce more azurophil granules. The implication is that phagocytosis plays a protective role in taking potentially dangerous material into the cell and that the azurophil granules destroy this foreign material using themselves up in the process. In the absence of significant phagocytosis substantial numbers of granules may accumulate in the mature cells and not become functional during the life of the cell. Like other examples cited, this situation may provide a reason for the consistent bounding of certain sorts of active substances within membranes. The substances are instantly available under certain situations, but they do not interfere with normal metabolism though being present for long periods.

Many Golgi apparatus appear to produce different materials at successive stages in development, but the Golgi apparatus in this instance appears to change the character of its product quite abruptly (Figs. 91 and 92).

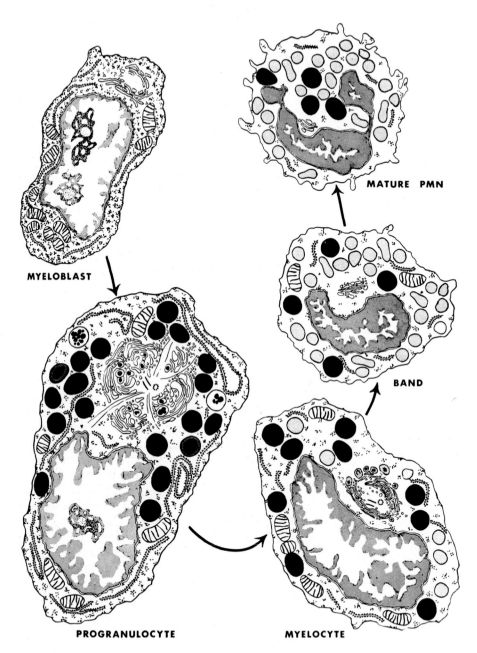

Fig. 90. A diagram of cellular changes with the development of a myeloblast to a mature polymorphonuclear leucocyte. From BAINTON and FARQUHAR, J. Cell Biol. **28**, 1966. Courtesy of Rockefeller University Press.

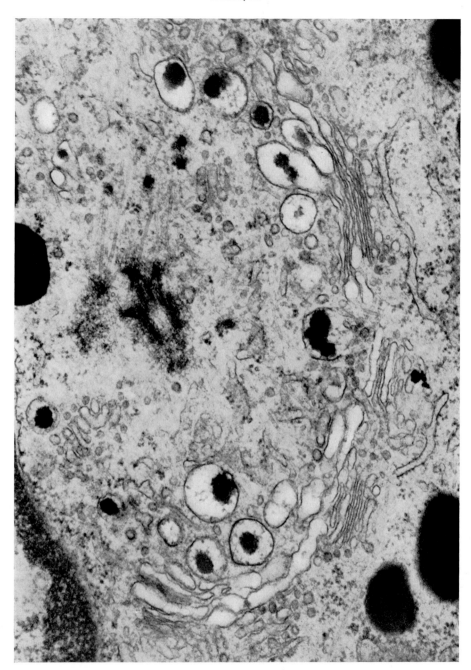

Fig. 91. Relatively early stage in the development of granules in the polymorphonuclear leucocyte. Successive stages of maturation can be seen beginning in the Golgi apparatus and ending with the dense granules at the edges of the micrograph. The association of the Golgi apparatus and the centriole is classic and was seen in light microscopy. From BAINTON and FARQUHAR, J. Cell Biol. **28**, 1966. Courtesy of Rockefeller University Press. ×44,000.

BAINTON and FARQUHAR (1966) have proposed that the azurophil granules are produced from one face of the Golgi apparatus and the later, differently staining granules, the specific granules, from the opposite face. This is difficult to ascertain with certainty since cell divisions intervene, but there is increasing evidence from quite different materials that there may be more than a one-way axis of activity in the Golgi apparatus. This is a clear example of the fact that the functional differentiation of the Golgi apparatus is a part of the functional differentiation of the cell.

BAINTON and FARQUHAR (1968 a, b) have identified peroxidase, acid phosphatase, arylsulfatase, β-galactosidase, β-glucuronidase, esterase, and 5'-nucleotidase in the azurophil granules of normal rabbit bone marrow. With the exception of peroxidase this complement of enzymes makes it most likely that the azurophil granules act as part of the lysosomal system. Leucocytes act in the destruction of foreign organisms that enter the bloodstream. For the most part such organisms are apparently engulfed by phagocytosis and destroyed by lysosomal enzymes characteristic of the leucocytes. Since the number of azurophil granules in any one leucocyte which act as lysosomes and participate in this destruction is limited, heavy infections are characterized by the increased production of leucocytes and the white blood cell count has become a diagnostic test for infection.

BAINTON and FARQUHAR (1970) have made a detailed study of enzyme distribution in rabbit eosinophils. The rationale of this study is based on the fact that these cells present a somewhat different relationship between enzyme production and secretion than has been worked out as an almost standard pattern by PALADE and his co-workers (see PALADE 1966). In these leucocytes, the protein production phase appears to be confined to the period when the differentiating cells are in bone marrow, and the protein synthesis mechanism appears to regress later when the mature, functional cells are circulating in the blood. At an early stage they found peroxidase, acid phosphatase, and arylsulfatase in rabbit eosinophils. These enzymes occurred in the perinuclear space, the rough endoplasmic reticulum, and the cisternae of the Golgi apparatus. With the progress of differentiation, acid phosphatase and arylsulfatase disappeared but peroxidase did not. BAINTON and FARQUHAR accepted their evidence as generally supporting PALADE's interpretation of the pathway of protein synthesis and movement though they commented that in the pancreatic exocrine cell, the proteins did not ordinarily move into the cisternae of the Golgi apparatus (there is now some question concerning this matter, see JAMIESON and PALADE 1971 b), whereas in this and some other instances they did. This instance may confirm the intracellular pathway of certain proteins but it does not extend the earlier observations of ARCHER and HIRSCH (1963 a, b) that eosinophilic leucocytes have a considerable appetite for antigen-antibody precipitates apparently as a consequence of lysosomal hydrolytic enzymes.

Considerable attention has also been given to the presence of mucosubstances in leucocytes. HARDIN and SPICER (1971) have made a detailed study of the mucosubstances in different types of rabbit leucocytes and they have commented at length that one of the functions of these mucosubstances

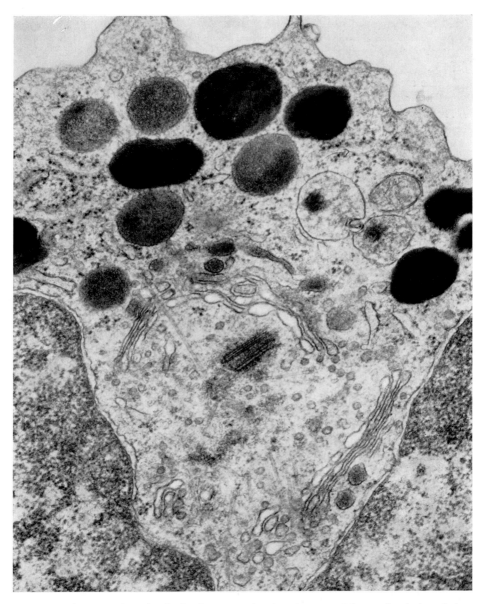

Fig. 92. A late stage in a developing leucocyte showing the azurophil granules dark and the specific granules light. From Bainton and Farquhar, J. Cell Biol. 28, 1966. Courtesy of Rockefeller University Press. ×44,000.

may be to inactivate the hydrolases which, as already noted, may remain in the cell for long periods of time.

Fedorko and Hirsch (1966) have followed the label of $^3$H-leucine in the granule production of heterophil-like leucocytes to arrive at one of the most clearcut distinctions between a passage of part of this label through the Golgi apparatus and the retention of part of it over the cytoplasmic matrix and the rough endoplasmic reticulum—thus distinguishing neatly between secretory and sedentary protein, a distinction discussed in detail by LEBLOND (1965).

There seems to be little doubt that the azurophil granules of certain leucocytes are part of a standing defense system prepared to destroy invading foreign material with the production of additional leucocytes providing a compensatory defense mechanism on the part of the organism. There is little doubt either that at least substantial amounts of any invading material are phagocytosed and destroyed within the leucocyte. Whether there is also a secretion of some of the hydrolytic enzymes involved in this protective mechanism is less clear. Equally unclear is the question of whether there are specific functions on the part of various types of leucocytes. Although a particular function has been established for the azurophil granules, less is known about the function of the specific granules. BAINTON (1973) suggests that the specific granules fuse with phagocytic vacuoles prior to the azurophil granules allowing the alkaline phosphatase, lactoferrin, and lysozyme contained within them to modify the engulfed material before its further digestion by lysosomal enzymes. She includes data suggesting differences in the membranes bounding the azurophil and specific granules (see also NACHMAN et al. 1972). This would be in keeping with the suggestion that functional differentiation of the Golgi apparatus results in changes in the characteristics of the membranes produced as well as in the enclosed products. It also implies that the membrane of the functional lysosome may be a mosaic membrane produced by the fusion of components of membrane systems that may differ significantly, i.e., plasma membrane, specific granule membrane, azurophil granule membrane. BAINTON also comments on the possible release of some of the granule enzymes during the phagocytic process (see also WRIGHT and MALAWISTA 1972). The functioning of the Golgi apparatus in at least some leucocytes appears to be unique. The character of the product of the organelle changes during differentiation of the cell. The early product, the azurophil granule, functions as part of the lysosomal system if it is called upon by stimulation by foreign material to do so. If it is not, it may, though made up of very active substances, persist through the life of the cell. If it does function, it does so in connection with phagocytosis and acts largely intracellularly. Here is a Golgi apparatus which provides protective substances but appears not to get involved in secretion in the usually understood sense.

## XI. Replication

One of the problems concerning the Golgi apparatus that the light microscopist thought he had solved despite the extreme pleiomorphism of the organelle and its variation with stage of development was the problem of replication. CAJAL, who in 1914 discussed the numerous developmental divergencies, considered in some detail the observations on this score of PERRONCITO (1910), MARCORA (1912), FAÑANÁS (1912), and DEINEKA (1912) all of whom believed some division of the structure was involved. There is also the earlier work already mentioned of replication of the organelle in the development of reproductive cells.

While all these investigators addressed themselves to changes in the structure during development, it was PERRONCITO who specifically explored the phenomenon of replication. He concluded (and his conclusions were substantiated by other investigators working with stages involving cell division) that there is a definitive division of the Golgi apparatus to some extent separate from the other events of cell division. PERRONCITO alone among GOLGI's students assumed a function for the organelle—the signaling of the onset of cell division. To this division of the organelle he applied, as noted, the term *dictyokinesis*. His concept was that upon entering dictyokinesis the Golgi apparatus became subdivided into smaller units to which he applied the term *dictyosomes*. The total evidence from the study of several investigators indicated that the distribution of the dictyosome might follow any of several patterns depending upon the material.

In one pattern, a portion—probably half of them—moved from one pole of the nucleus to the other, thus assuring equivalent distribution on cytokinesis. Other patterns involved the grouping of dictyosomes around the equatorial region of the cell or their random distribution, both patterns by some mechanism resulting in equivalent distribution at cytokinesis. Whatever the pattern, it included fragmentation of the organelle. The relative discontinuity exhibited by many Golgi apparatus in electron micrographs would make it very difficult to follow such processes in detail. One is obliged to consider other possibilities not only for its replication but also for its origin.

As is true of nearly all cellular organelles, it has occasionally been proposed that the Golgi apparatus arises *de novo*. This suggestion was apparently made by VON BERGEN as early as 1904 and it was repeated by BRAMBELL and LUDFORD separately in 1925. *De novo* origin of organelles has never been a popular hypothesis and proposing it has frequently led to definitive studies of replication. There are some valid reasons for the negative reactions to the question of *de novo* origin of organelles. Continuing coordinated functioning of cells depends heavily upon genetically controlled continuity in cell structure. Even if this were not so *de novo* origin poses an unanswerable question: do specific regions of the cytoplasm have attributes for development into an organelle like the Golgi apparatus but lack the normal form and function of the organelle or does differentiation impart such attributes to specific regions? If so, what is the basis for this development in a particular region?

There are, however, still some instances of apparent *de novo* development of organelles which raise questions. One is the origin of the centrioles in certain lower plants which are without centrioles during their vegetative phases, but exhibit such organelles during reproductive phases. Replication of centrioles during cell division even in such instances seems to involve specific spatial arrangements though its mechanism is not clearly understood. In the state of knowledge at the time, WILSON (1925) could do no more than conclude that the Golgi apparatus "is in some sense self-perpetuating". He left little room for serious consideration of *de novo* origin.

BOWEN (1929) came to almost the same position, concluding that the substance of the Golgi apparatus had some unknown means of replicating itself and that the mechanism bore no relation to karyokinesis. He, too, gave short shrift to the *de novo* origin hypothesis. However, the idea is still around and in some senses is harder to explain away than it was earlier.

The formalities of PERRONCITO's concept of dictyokinesis fit much better the early observations on the Golgi apparatus as a network observed to be in association with the centrioles. Even WILSON recognized both a fragmentation and a movement in association with the centrioles sometimes to be conspicuously involved in the mechanics of cell division. The later observations by electron microscopy that the organelle is more discontinuous has made fission an attractive possibility, as it has also made development at organizational centers.

GATENBY had proposed reproduction by fission as early as 1919; GRASSÉ (1957) proposed it from electron micrographs. There have been several other proposals of division by fragmentation, some actually purporting to show in which cisternae the process begins. It may be that the organelle divides differently in different types of cells or stages of development, but it is a complex structure involving differentiation from face to face and from the center to the periphery and its reproduction seems likely to involve some complex principles of cellular organization and not to be simply a hit or miss fragmentation.

In many Golgi apparatus individual cisternae appear variously split and fused. This has led to a common assumption that the apparatus may split and that this splitting may start in an individual cisterna and progress with continued activity of the organelle. Micrographs showing Golgi apparatus side by side, which could possibly represent the results of splitting, are fairly common. In MIGNOT's (Fig. 23) micrographs showing intercisternal material, it may occur at intervals giving a picture of an organelle which could replicate by splitting under certain circumstances. To an extent this question of replication by splitting is tied up with the question of whether there is a displacement of the cisternae across the stack.

Although fragmentation has probably been the most popular hypothesis, alternate suggestions have been made (for review see WHALEY 1966, MORRÉ *et al.* 1971). These include development of the membranes of the apparatus from aggregations of nonmembranous materials, the "growth" of a single Golgi cisterna, and derivation from other membrane systems within the cell. All of these possibilities might reflect on the concept of a Golgi region and activities which could take place within such a region. Some light can be shed on the contributions from other membrane systems from studies directed primarily toward Golgi apparatus functioning.

HALL and WITKUS (1964) and WHALEY *et al.* (1964) have shown that under certain circumstances additional cisternae can be formed on one face of the Golgi apparatus. NEUTRA and LEBLOND (1966 a) have proposed the normal formation of such cisternae as compensation for the transformation of those at the opposite face into mucigen-containing vesicles in goblet cells. BROWN (1969) and MOLLENHAUER (1971) have proposed a similar mechanism in other types of cells. This process involves the organization of cisternal envelopes with membranes either being constituted at one face of the Golgi apparatus or transferred from elsewhere in the cell. It provides evidence that the formation of the membranes of the Golgi apparatus is a result of processes semiseparate from the extension of membranes in the organelle.

It is a common observation that in electron micrographs of lower organisms or early stages of development blebs from the nuclear envelope appear to be moving toward the proximal face of the Golgi apparatus (Fig. 34). It is an equally common observation that in other sorts of material that membrane-bounded vesicles seem to be moving from the endoplasmic reticulum to the Golgi apparatus (Fig. 33).

Many investigators have supposed that such membrane-bounded vesicles fuse to form the more proximal cisternae of the apparatus to which other cisternae are added in the normal course of development. Some very difficult interpretations are involved in this problem. One is that the size of the vesicles and the widths of the proximalmost cisterna(e) are very disparate and there would have to be some very rapid changes involved. Perhaps more important still are the demonstrations from PALADE's laboratory (MELDOLESI et al. 1971 a, b, c) that intracellular membrane systems are notably different in composition. These investigators question directly the possible transfer from other cellular membrane systems to the Golgi apparatus, though there is considerable evidence that there may be such transfer accompanied by rapid changes (see WHALEY et al. 1971).

BOUCK (1965) has proposed that in the brown algae the nuclear envelope is the source of the most proximal cisterna(e), and KESSEL (1971) has postulated that in grasshopper embryogenesis the whole Golgi apparatus is derived from transfers of nuclear envelope. CHRÉTIEN (1972) has followed restoration of function in the submaxillary gland of the castrated male mouse after testosterone injection. She suggests that at first new Golgi apparatus are produced by activities of the nuclear envelope and that subsequently the rough endoplasmic reticulum also contributes to the developing apparatus. Interrelations between the nuclear envelope and the endoplasmic reticulum have been cited repeatedly. There is some evidence (PARKS 1962) that the endoplasmic reticulum is initially derived developmentally from the nuclear envelope. This would tend to support KESSEL's (1973) contention that the nuclear envelope is a substantial producer of intracellular membrane.

All these speculations leave unanswered the question of whether the Golgi apparatus exists at certain stages of cell development in forms other than those which we recognize. There is increasing evidence that there may be alternate forms of the organelle characteristic of certain stages of cellular activity or relative inactivity and that these alternate forms (most commonly suggested are aggregations of small vesicles) may be reorganized periodically into the usual Golgi stacks as the functions of the cells change (OVTRACHT 1972, SANDOZ 1972). It even appears that alternate forms may be experimentally induced (see Section XII). Although unrecognizable forms of the organelle might be involved, GOODMAN and RUSCH (1970) have observed what appears to be an absence of Golgi apparatus in the plasmodial cells of myxomycetes. When such cells are put under conditions conducive to sporulation, distinctive Golgi apparatus appear. Embryos in dry seeds of *Zea mays* appear to lack characteristic Golgi apparatus but such are apparent when the seeds are germinated (see also YOO 1970). Such observations as these invite speculations that specific attributes of membrane building and membrane specialization in the Golgi apparatus may be essential to the morphogenetic development of organisms.

Another observation concerning replication is worthy of note. WALNE (1967) discovered that when cells of *Chlamydomonas* are put into colchicine solutions the one or two Golgi apparatus present seem to increase in number parallel to the buildup of polyploidy and then undergo a reduction upon recovery of the cells.

FLICKINGER (1968 a, 1969) performed a series of experiments in which amoebae were enucleated and then renucleated after a period of time. The enucleated cells showed a diminution of the Golgi apparatus but the organelle returned to its usual form when the cells were renucleated.

The removal and transfer of nuclei without serious contamination leaves questions concerning such experiments, but the general evidence seems to favor some association between the nucleus and its contents and the number and stage of development of the Golgi apparatus. In further experiments, utilizing inhibitors of protein and RNA synthesis, FLICKINGER (1968 b, 1971 a, b) concluded that total reconstitution of the Golgi apparatus depended upon certain specific RNA activities. This is hardly surprising since a primary function of the organelle is membrane building and this function requires an ample supply of specific proteins.

BEAMS and KESSEL (1968) cite some other possibilities as to origin. MERCER (1962) proposed, for example, that the organelle arises by a series of conversions of stored phospholipid. DANIELS (1964) suggested that it is formed from plasmalemma vesicles which originate by pinocytosis or phagocytosis. MERCER's proposal involves an inordinately complex series of events and has to do with only one constituent of the system. DANIELS' plasmalemma vesicles are just as likely to be on their way to the cell surface as moving from it to the Golgi apparatus.

Despite our ignorance of the mechanism involved in replication or of whether there is more than one mechanism it is possible to point out certain pertinent facts about replication. It may or may not relate to cell division. In the cells of maize rootcap just before the onset of the intense secretory phase, there is a considerable increase in the number of Golgi apparatus. These cap cells are beyond the division stage, yet the Golgi apparatus increase and enlarge until they are by far the most conspicuous feature of the cytoplasmic landscape.

In the case of most other organelles, fission or some more complex process has proved to be the mechanism of replication. For Golgi apparatus the mechanism of replication is still unknown. It may involve a series of changes in morphology, but it depends upon a supply of lipids and proteins sufficient to form large amounts of membranes. One of the activities of the genome in controlling the ontogeny of cell types most likely involves the specialization (probably internal as well as surface-associated) of these membranes.

Not many other aspects of the Golgi apparatus and its functioning are so confusing as the evidence that bears upon its origin or its replication. There is no certainty that a single process is involved or even that the organelle has been seen in the form or the sequence of forms in which replication takes place. Behind the whole question is a concern with whether or not there are present in the cytoplasm conditions that create a situation favorable for the organization and the specialization of the membranes or whether the cell has the capacity to fragment by some mechanism an organelle characterized by various axes of activity. The one fact made clear by both the earlier work and the recent studies is that the earlier ontogenetic stages of an organism appear to be characterized by the structurally seemingly simple dictyosomelike form of the apparatus. In some cells differentiation involves modification of this form into extensive groups of cisternae, sometimes in very specific arrangements; in other cells the morphology of the simple form does not seem to be substantially altered with cellular development, and the simple form continues to be apparent in some highly complex cells including vertebrate neurons. The complexity of form seems to be unrelated to the ability of the apparatus to function as an organelle.

## XII. Modifications and Dysfunctions

Modifications of the Golgi apparatus are difficult to deal with because of the many, varied forms of the organelle, its extensive changes with activities of the cell, and significant alterations during the period of development and differentiation.

There is ample evidence that it responds with alterations, morphological and compositional, to various experimental treatments, and in relation to dysfunctions.

At a very fundamental level modifications of the Golgi apparatus and its functioning are of concern in relation to the congenital lysosomal diseases mentioned earlier. A few other clearcut instances which relate the organelle to genetic dysfunctions are beginning to emerge. SANYAL and BAL (1973) have related the degeneration of photoreceptor cells in mice homozygous for the mutant gene rd/rd to distinctive modifications in the Golgi apparatus which is known to play a critical part in retinal development. They furthermore related the early disruptions and degeneration of the Golgi apparatus to altered surface properties of the cells. SEEGMILLER et al. (1971, 1972 a) have shown modification of the Golgi apparatus to be associated with a chondrodystropic mutation in mice. PLATZER and GLUECKSOHN-WAELSCH (1972) have suggested that a mutationally controlled muscular dysgenesis in mice is associated with cellular changes in several organelles including the possibility of the Golgi apparatus being encompassed within the nuclei. This is of particular interest in relation to material discussed later in this work because they conclude that the mutational effect may be related to a disruption in normal membrane development.

LANDIS (1973) has worked out the nature of the ultrastructural changes in the Purkinje cells of homozygous nervous mice (*nr/nr*). In all but a few cells there are extensive changes in the mitochondria, the general form of the endoplasmic reticulum and polysome association with it, and the Golgi apparatus. This example illustrates the interrelationships of a number of cell components in carrying to completion essential metabolic functions. Ultimately there is complete degeneration of the cell, but before this final stage is reached, LANDIS notes that the cisternae of the Golgi apparatus appear to fragment, scatter in clusters of vesicles, then disappear. It may be that an essential function or substrate is lost prior to this development. One of the other characteristics of the degenerative process is a general reduction in membrane systems.

The list of substances that will affect the activities of the Golgi apparatus is too long to permit anything but very selective review. They range all the way from pilocarpine which CAJAL used in his 1914 experiment to stimulate secretion of certain glands to various adrenergic blocks which are useful in reducing the production or the release of catecholamines.

In plant cells the introduction of 6 aza uracil by HALL and WITKUS (1964) and the lowering of oxygen availability or the introduction of potassium cyanide by WHALEY et al. (1964) both interfed with the transport of secretory products but allowed a continued buildup of cisternae. FLICKINGER (1972) found that in enucleated amoebae which had been renucleated, cyanide and dinitrophenol both limited the redevelopment of the Golgi apparatus. Fluoride was less effective. These results led FLICKINGER to conclude that mitochondrial oxidative phosphorylation was involved in the rebuilding of the Golgi apparatus.

FLICKINGER (1971 b) also exposed amoebae to emetine. The Golgi apparatus first enlarged and then decreased in size and number. The effect was somewhat different than that obtained with actinomycin or other protein inhibitors. FLICKINGER discusses the implications of the inhibition of protein synthesis for what he terms the normal turnover of components of the organelle, but both the membranes and the central cores of most secretory products contain protein and one would expect some effect from such treatments.

WHETSELL and BUNGE (1969) found that when they added ouabain to cultures of rat and mouse sensory ganglia there was first a swelling of many but not all the Golgi apparatus. Then nearly all the Golgi apparatus disappeared and there appeared in their place large numbers of vacuoles. Ouabain is known to interfere with sodium and potassium transport. When the ouabain was removed from the cultures, what appeared to be normal Golgi apparatus reappeared. This experiment not only suggests that the organelle may play a part in sodium and potassium transport, but it also suggests that it may be possible under experimental conditions deliberately to create an alternate form of the Golgi apparatus.

In a rather complex experiment aimed at studying the synthesis and movement of proteins in the rat somatotroph HOWELL and WHITFIELD (1973) came to the conclusion that inhibitors of protein synthesis did not inhibit the transfer of already synthesized protein from the endoplasmic reticulum to the Golgi apparatus (as had been shown for the pancreas, see Section VI, B) nor did the inhibition of Na-K-dependent ATPase by ouabain. Dinitrophenol which inhibits oxidative phosphorylation and antimycin A both reduced ATP levels and inhibited the transfer. From their total evidence they concluded that ATP is essential for the transfer and also plays some part in the formation of storage granules. Both dinitrophenol and antimycin A caused dilations in the endoplasmic reticulum and disorganization of the Golgi apparatus (Fig. 93). They assumed from this experiment which involved studying the distribution of label from ³H-leucine that the transfer takes place in what they characterize as "transfer vesicles".

Melanin is formed by a series of several steps from tyrosine; the conversion is catalyzed by tyrosinase which is found in vesicles associated with the Golgi apparatus and in GERL (for references see WRATHALL et al. 1973). In an experiment with a highly pigmented clone of mouse melanoma cells, WRATHALL et al. found that exposure for extended periods to 5-bromodeoxy-uridine (BrdU) would completely suppress the formation of pigment granules (Fig. 94). The activity appeared to be through the suppressing of the synthesis of tyrosinase and thus modification of an enzymatic characteristic of the Golgi apparatus. For further information on the effects of BrdU see HOLTZER et al. (1973) and RUTTER et al. (1973).

SEEGMILLER et al. (1972 b) introduced 6-AN into chick embryos. They found that as a result the rough endoplasmic reticulum was greatly reduced and the Golgi vesicles that normally transport mucopolysaccharides were small or absent (Fig. 95). They hypothesize that the result was a deficient matrix. They suppose tropocollagen to go into this matrix via the Golgi apparatus. In the presence of 6-AN there was very little fibrillogenesis in this matrix, and the paucity of intercellular supporting material ultimately led to death.

That the Golgi apparatus is modified by many different sorts of injury effects is not only indicated by the sort of evidence presented here but also was suggested as early as CAJAL's 1914 observation that it is a particularly vulnerable organelle in cell death. If the cell is able to compensate for the injury imposed, the organelle will ordinarily recover its original characteristics. A simple example will serve to indicate a response to physical injury. RAJARAMAN and KAMRA (1969) exposed cells of Ulva lactuca, an alga, to two different types of laser beams. They found the greatest damage to be caused to the chloroplasts and the Golgi apparatus (Fig. 96). A ruby laser caused irreparable cell damage. Damage by a neodymium laser was less extensive. The authors comment on the particular sensitivity of the Golgi apparatus.

PRICE and PORTER (1972) studied the recovery of ventral horn neurons of the frog following axonal transection. A result was degranulation of the endoplasmic reticulum and some conspicuous changes in various components of the cell including the Golgi apparatus and its association with the lyso-

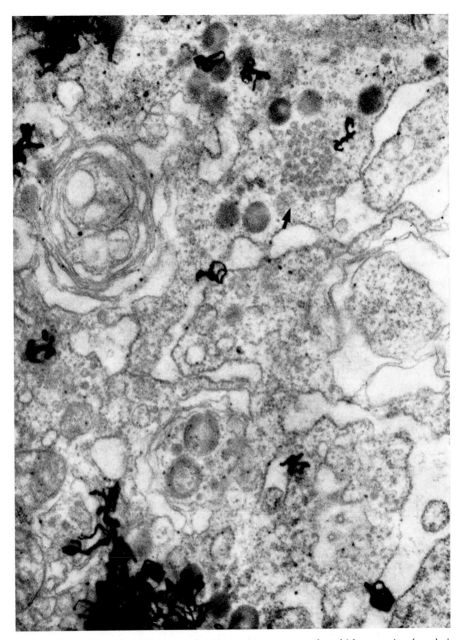

Fig. 93. A $^3$H-leucine radioautograph of a rat somatotroph which was incubated in 2,4-dinitrophenol. The endoplasmic reticulum is distended. What the authors interpret as transfer vesicles (arrow) are numerous and the Golgi apparatus looks abnormal. From HOWELL and WHITFIELD, J. Cell Sci. 12, 1973. Courtesy of Cambridge University Press. ×24,700.

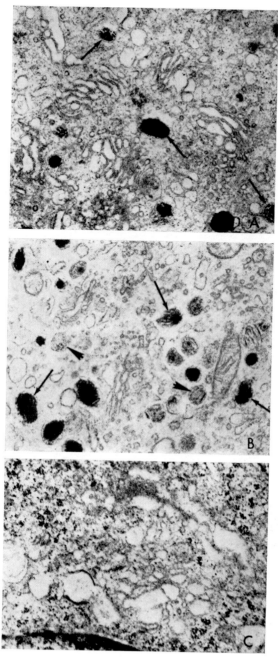

Fig. 94. B 16 mouse melanoma cells. *A.* Control. Normal pre-melanosomes are seen at arrows. *B.* Up to three days exposure to BrdU. There appears to be an increase in the pre-melanosomes in this area. *C.* Seven days exposure to BrdU. No pre-melanosomes are seen in the vicinity of the Golgi apparatus as in the other micrographs. From WRATHALL *et al.*, J. Cell Biol. **57**, 1973. Courtesy of Rockefeller University Press. *A.* ×20,500, *B.* ×20,000, *C.* ×28,000.

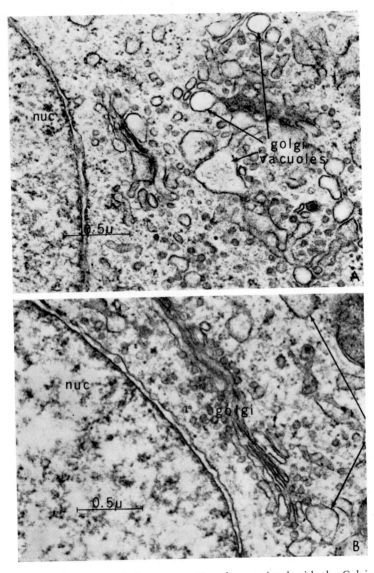

Fig. 95. *A*. Normal six-day old chondrocyte. Vacuoles associated with the Golgi apparatus are believed to contain chondroitin and tropocollagen. *B*. Six-day-old chondrogenic cell after treatment with 6-AN. There are few conspicuous vacuoles associated with the Golgi apparatus, and the treatment prevents any involvement of a substantial intercellular matrix. Arrows indicate rough endoplasmic reticulum. From SEEGMILLER *et al.*, Dev. Biol. **28**, 1972 b. *A*. ×34,000, *B*. ×34,000.

somes. They followed the material through a fairly long postoperative period to demonstrate that some of the organelle changes represented different phases of the recovery process.

Although it is not possible at this stage to provide a satisfactory interpretation of the replication of the Golgi apparatus or a detailed evaluation

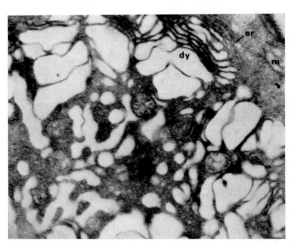

Fig. 96. Distinct hypertrophy of the Golgi apparatus after exposure to a ruby laser beam. Cell of *Ulva lactuca*. *dy* = Golgi apparatus cisternae. *er* = endoplasmic reticulum. *m* = mitochondrion. From RAJARAMAN and KAMRA, J. Ultrastruct. Res. **29**, 1969. ×27,000.

of its response to abnormal modifications of metabolism, it does appear that the cell is capable under normal circumstances of providing enough apparatus to meet its requirements in development and functioning. If these requirements turn, as suggested in the concluding section, around final stages in the assembly of informational macromolecules for transport to the surface of the cell, then the fact that the organelle appears to be very sensitive to modification may be of considerable consequence, for there is increasing evidence that many of the cell's characteristics are dependent on a system of surface-associated determinants. So far the experimentally induced alterations have involved rather general aspects of cellular metabolism. It may be possible with techniques yet to be developed to systematically modify the cellular organelles controlling aspects of development and dysfunctioning. The findings presented lend credence to the idea that much genetically controlled development and functioning is mediated by this organelle.

## XIII. Concluding Remarks

The early studies suggested the presence of the Golgi apparatus in a wide variety of cell types. Further studies contributed an association of it with secretion and the formation of the acrosome. Actually they made it a sort of focal point in the final assembly of secretory materials.

Improved ultrastructural and biochemical techniques have added many of the details in the assembly process. This process begins with the coding of polypeptide chains in the ribosomes with also the possible addition of some carbohydrate moieties. Proteins are polymerized in the endoplasmic reticulum often accompanied by further addition of sugars, and the final assembly, sometimes with the addition of still more sugar groups, occurs in the Golgi apparatus or its derivatives. Thus one has the progressive assembly of relatively complex conjugated compounds by the cooperative but separable activities of the endoplasmic reticulum and the Golgi apparatus. Formation of the final storage form of the material may involve further modification. For example, some proteins (such as many of the enzyme precursors in pancreatic zymogen) do not have added carbohydrate moieties, but carbohydrates added to the secretion vesicles by the Golgi apparatus may affect their storage prior to secretion. In other cases proteins may be modified enzymatically in the apparatus. The synthesis and secretion of insulin in the endocrine pancreas has been shown to occur by the usual pathway (ORCI et al. 1971, 1973 b). The precursor form of insulin (proinsulin) is found in the endoplasmic reticulum, and it has been suggested that the conversion to insulin which involves the tryptic removal of a polypeptide portion of the molecule may occur in the Golgi apparatus (HOWELL and LACY 1971, GRANT et al. 1971). Either enzymatic activities or the influence of carbohydrate material may be involved in the changes described by WEINSTOCK and LEBLOND (1974) in the odontoblast. Other compounds, for example, metals (see GRANT et al. 1971) and nucleotides (see HOWELL and EWART 1973) may facilitate the final formation and stabilization of secretory materials in the Golgi apparatus or in the derived secretory vesicles. The high degree of morphological structuring often seen in these vesicles also implies a complexity of function in the Golgi apparatus about which more information is needed.

More specific data are also needed on the process of vesicle exocytosis. In a number of instances it seems reasonably clear that microtubules may act in guiding the vesicles to the proper position for exocytosis; however, membrane fusion may require in addition some sort of recognition phenomena and further stimuli for secretion. A possible role for microfilaments has also been suggested (for example ORCI et al. 1972). The question of the extent to which exocytosis and endocytosis may be coupled (DOUGLAS et al. 1971, MASUR et al. 1971, 1972, ORCI et al. 1973 a) provides a most interesting area for further research as does the question of how activities in the Golgi apparatus may affect these phenomena.

The materials produced and secreted differ from one tissue type to another, often from one cell type to another, and perhaps even from one Golgi apparatus to another—certainly from one period of activity to another. But they have in common certain characteristics that are deemed to be of importance for the interpretation of the functioning of the organelle. They are relatively large structures with a substantial amount of complexity, and many of the molecules of which they are composed including several of the sugar groups that are thought to be attached in the Golgi apparatus are materials known from other sorts of studies (see AMINOFF 1970) to have high degrees of specificity. It has been suggested (see DAUWALDER et al. 1972, WHALEY et al. 1972) that macromolecules of this sort are unusually good carriers of biological information. The proportions of the different components may differ substantially from one product to another, and the specificity that they carry may in one instance lie in a terminal carbohydrate group and in another instance in the sequence of amino acids exposed after the carbohydrate side chains have been removed. One example of the effects of altering the terminal sugar sequence is seen in the enhanced uptake of circulating glycoproteins by the liver following removal of the terminal sialic acid groups to expose galactose groups and the subsequent return to a normal rate of uptake with the further removal of the exposed galactose residues (see GORDON 1973). Other examples of the importance of the structuring of carbohydrate moieties are seen in substances with blood group activity (AMINOFF 1970) and in the "antifreeze" property of blood glycoproteins in several species of Antarctic fish (see MARSHALL 1972); however, for the majority of glycoproteins possessing biological activity there is not yet a clear understanding of the precise functional contribution of the carbohydrate groups. From what is known it seems reasonable to predict that an increasing number of biologically important functional specificities will be linked to the carbohydrate moieties of secretory materials and that in many instances activities in the Golgi apparatus will be shown to be involved in their synthesis.

The emphasis on the importance of the Golgi apparatus in secretion as classically defined though obviously of critical importance for the integrated functioning of the organism may not be as significant as supposed for a general consideration of the functioning of the organelle. It now appears that in many if not all cell types the Golgi apparatus may be implicated in the turnover of the cell surface, and it has been suggested that functional activities within the organelle may be responsible for providing components of the surface membrane and in the structuring of carbohydrate groups which confer specificity characteristics on the surface materials (WHALEY et al. 1972, DAUWALDER et al. 1972). Study of materials coating cell surfaces was advanced by CHAMBERS (see CHAMBERS 1940) who suggested that in some specialized cell types various physiological properties might belong to extraneous coatings rather than the actual protoplasmic surface of the cell. H. S. BENNETT (1963) suggested tentatively that a carbohydrate-rich surface might exist on all cells. Major contributions to the general occurrence of carbohydrate-containing cell "coats" as well as their possible origin from the Golgi apparatus have come from LEBLOND and his co-workers (RAMBOURG et al. 1966, RAMBOURG and LEBLOND 1967, RAMBOURG et al. 1969, G. BENNETT 1970, G. BENNETT and LEBLOND 1970, RAMBOURG 1971, G. BENNETT et al. 1974).

Interest in and study of the cell surface has become widespread (for recent reviews see BURGER 1971 a, b, HAKOMORI 1971, KRAEMER 1971, SCHMITT 1971, D. BENNETT et al. 1972, OSEROFF et al. 1973) and has attracted the attention of immunologists, geneticists, biochemists, and members of the various

disciplines of cellular biology. Surface groups have been implicated in growth
regulation, mitotic regulation, malignant transformation, embryogenesis and
morphogenesis, cellular recognition, cellular antigenicity, and as receptors
for hormones and other active agents (the list is not complete). Of particular
concern here is the clearly demonstrated importance of carbohydrate-
containing groups in many of these phenomena, how modification of these
groups changes the behaviour of cells, and the extent to which activities in
the Golgi apparatus may be implicated in the "control" of specificity characteristics of the surface. Differences in methodology and approach make it
difficult to establish clear correlations in much of the above information,
but the interrelations among the data have allowed the development of a
working model. The evidence that the Golgi apparatus is important in the
transport of materials out of the cell; that Golgi vesicle membranes are
incorporated into the plasma membrane; that the Golgi apparatus is involved
in the synthesis of polysaccharides and in the addition of the more terminal
sugars of glycoproteins; that glycoproteins are involved in functions at the
surface and that this involvement can in some cases be linked to the carbohydrate moieties and in a few cases even to the particular sugar sequences;
and that in some cases the sequence of sugars is under strict genetic control
have all led to a consideration of the Golgi apparatus as a discrete site in the
genetically controlled assembly of macromolecules which come to confer
informational characteristics on cell surfaces (DAUWALDER et al. 1972).

As noted there is evidence implicating the Golgi apparatus in the synthesis
and transport of cell coat materials (for information and further references
see the papers from LEBLOND's laboratory); however, clear evidence linking
specific synthetic mechanisms in the Golgi apparatus and specificity groups
on the surface is lacking. Of considerable importance here is the question
of how such synthetic mechanisms could be genetically regulated and the
nature of the coordinated interactions of the genome, the Golgi apparatus,
and the cell surface (see BENNETT et al. 1972). As noted by KRAEMER (1971)
some mechanisms involving RNA have been suggested. However, the most
acceptable current thinking is that groups of glycosyltransferases operate
cooperatively in the biosynthesis of complex carbohydrate moieties. None
the less, some sort of activities limited to the functioning of the Golgi apparatus must be involved, for the type of sugar addition which can occur
in the endoplasmic reticulum is distinctive from that occurring in the Golgi
apparatus. Such complexes of enzymes must either be added to, activated in,
or perhaps located in a proper spatial distribution, or combined with necessary coenzymes within this organelle. This brings us to the concept that one
of the primary functions accomplished within the apparatus is the specialization of membranes, and that the assembly of various products might be
viewed as a consequence of this specialization (see SJÖSTRAND 1968). Considering the morphology of the organelle, its capacity to add cisternae, and
the frequently localized buildup of secretory products, this concept carries
with it a difficult-to-understand proposition of substantial genetically controlled differentiation within the organelle itself.

Despite concerted effort on the part of many researchers, many aspects of membrane structure, function, and biosynthesis are still unclear. SIEKEVITZ (1972) has interpreted cellular membranes as having enough structure to lend them integrity but enough openness to allow the dynamic turnover of structural components and the passage of various sorts of molecules. These membrane systems must also have sufficient integrity to maintain a great deal of differentiation both within membrane systems and probably within given regions of individual membranes. The Golgi apparatus provides a prime example of membrane modification. Differences between the membranes of the apparatus and those of the endoplasmic reticulum have been noted; and furthermore differences within the apparatus itself have been shown by morphology (see HICKS 1966, GROVE et al. 1968, HELMINEN and ERICSSON 1968 a, STAEHELIN and KIERMAYER 1970), by staining of products (see RAMBOURG et al. 1969, RAMBOURG 1971, OVTRACHT and THIÉRY 1972), by cytochemical localization of enzymes (see NOVIKOFF et al. 1971), by impregnation techniques (see DAUWALDER and WHALEY 1973), and by the biochemical or cytochemical analysis of isolated fractions (see MELDOLESI and COVA 1972 a, BERGERON et al. 1973, FARQUHAR et al. 1974). Such differences may be exhibited in changes occurring unidirectionally across the Golgi stack, bidirectionally within the stack, within individual cisternae of the stack, or within regions of a single cisterna. Hypothetically, at least, the functional differentiation of the Golgi apparatus involves specific changes in the cisternal membranes as well as in their associations with enzymes responsible for the synthesis of compounds with a great deal of biological activity. The ontogenic, physiological, and functional specificity of this differentiation implies again strong influences of the genome.

The manufacture of surface materials would provide a function for the Golgi apparatus in cells generally. The possibility of the assembly of these surface materials, units of plasma membrane with associated specificity characteristics, would give the activities of this organelle a focal role in cellular development, specificity, and association with the differentiation, functioning, or malfunctioning of multicellular organisms. The involvement of the cell surface in activities of this sort is already well established as a determining factor in the sociology of many unicellular and colonial organisms. This concept also suggests that not only must the Golgi apparatus be looked upon as an organelle, but the membrane itself must be considered to have many of the characteristics of an organelle in the sense that it becomes highly specialized in vastly different ways in response both to modifications in the genome and to experimental conditions. The concept makes secretions in whatever specialized form a consequential function of the particular genetic control of the membrane development. It also obviously calls for further investigation of the character and role of cellular membranes.

Material which is both structural and informational may be transferred to the cell surface at a rate that conforms to the requirements of activities of this surface for change. This change is in turn related to the intensity of intracellular events and events which take place extracellularly as part of the normal development and functioning of the organism.

One of the observations of electron microscopy has been the frequency with which certain cell surfaces carry out various endocytic activities which may provide nutrient for cell function, may allow for modification of intercellular matrix materials, and may also include masses of surface-associated material that have served their particular function. At least much such material is membrane-bounded and within the cell it fuses with the active hydrolases of the lysosomal system which, to the extent they have been clearly identified, are products of the Golgi apparatus. The result is a return of material to pools for reutilization in the cytoplasm. There is evidence that this process is correlated with the synthesis of new materials particularly those that control development and activities through the continuing activation of portions of the genome. Problems of senescence and breakdown which may be local even if reflected systemically may have to do with the balance between continuing synthesis and reutilization. It is attractive to suppose that in any system as fundamental as the cellular membrane system there is not only some concern for recycling but there is a progressive change with the cell surface and its associated specificities being the ultimate site at which the information assembled finds certain degrees of expression.

A substantial body of evidence that the Golgi apparatus has a central function in both the assembling of the membrane and the informational specialization is summed up in DAUWALDER et al. (1972). Elements of this hypothesis have been presented by numerous investigators in relation to the activity of the Golgi apparatus in certain types of cells (for example see DINGLE 1969) (Fig. 97). If one applies this sort of an interpretation to the functioning of the Golgi apparatus one gives the organelle a central position in cellular development and functioning while retaining the specific importance of its other activities in various types of differentiated cells. In this central position there must necessarily be additional considerations of the processes involved in the passage of water, ions, and molecules and of both the maintenance of balances and the possibility of compensatory effects induced by cellular responses to specific conditions.

The steps in the membrane specialization and those involved in the assembly of informational macromolecules offer a reasonable explanation of the mechanism of genetic control over surface characteristics. These surface characteristics are modified both by environmental events and by intracellular processes. This more generalized functioning of the Golgi apparatus is still too much of a postulate to justify a long cataloging of the influence of informational molecules, their specificities, and their possibilities for combining in various ways to contribute to specific phases of development, but it seems worthwhile to mention a few examples in which conjugated components synthesized and assembled via this system may function.

Studies on various aspects of cell recognition and aggregation were actually initiated long ago with the findings of WILSON (1907) that the reaggregation of dissociated sponge cells was species-specific. Current research along these lines has led to a clearer understanding of the surface groups involved and suggested that such compounds might present a key to many problems of development and morphogenesis. KLEINSCHUSTER and MOSCONA (1972) have shown in retinal development that changes in carbohydrate-containing surface groups of embryonic cells were correlated with differentiation and maturation. Additional information has been reviewed by MOSCONA (1971, 1973).

The association of similar mating types of *Chlamydomonas* and other algae has been shown to involve recognition by specific glycoproteins and

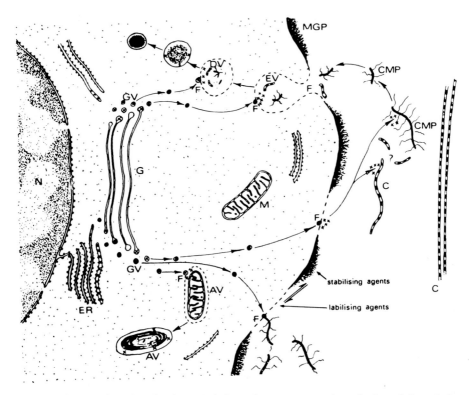

Fig. 97. Diagram depicting the dynamic balance between rates of synthesis and degradation in a cell in relation to its environment. The possible pathways of Golgi apparatus-produced vesicles in the secretion of polysaccharide matrix components, in the secretion of lysosomal enzymes, or in the addition of lysosomal enzymes to endocytotic vacuoles are shown. *MGP* = membrane glycoprotein, *EV* = endocytotic vacuole, *DV* = digestive vacuole, *GV* = Golgi vesicles, *G* = Golgi, *N* = nucleus, *ER* = endoplasmic reticulum, *M* = mitochondrion, *AV* = autophagic vacuole, *C* = collagen, *CMP* = chondromucoprotein, *F* = points of membrane fusion, ● = Golgi vesicles containing polysaccharide, X = Golgi vesicles containing lysosomal enzymes (primary lysosomes), – – – – = physicochemically altered membrane lipids, XXX = active lysosomal enzymes. From DINGLE, in: Lysosomes in biology and pathology, Vol. 2 (DINGLE, J. T., and H. B. FELL, eds.). Amsterdam: North-Holland. 1969. Also used, with permission, in DAUWALDER *et al.*, Sub-Cell. Biochem. 1, 1972.

it may be that much of the association of such organisms is dependent on this sort of activity. W. A. ANDERSON (1968) has shown evidence of recognition relationships between the egg and sperm of the sea urchin, and it may be that factors of this sort are involved in mating relationships among numerous species. Other activities of the Golgi apparatus are involved in other aspects of reproduction (E. ANDERSON 1968, BEAMS and KESSEL 1969, BACCETTI 1970, SUSI et al. 1971, COGGESHALL 1972 b, SCHUEL et al. 1973, DEWEL and CLARK 1974).

The completion of immunoglobulins in the Golgi apparatus and their specificities as antibodies suggest that many defense mechanisms of the organism are mediated by this organelle. Of primary concern also is the compatibility of grafts and other purposely undertaken reassociations of cells. These too appear to depend on the operation of this assembly system and the expression of genetic control through it. Some examples of histocompatibility have been achieved by suppressing certain characteristics of the immunological system but doing so presents great dangers related to the lack of resistance. A diligent search for further means of modifying the systems as a consequence of affecting the assembly and transfer might well provide better answers to a whole range of difficult problems. (The Golgi apparatus is also important in some quite different defense mechanisms SLAUTTERBACK 1963, BOUCK and SWEENEY 1966, SKAER 1973.)

Much attention has attached to the modification of the genome of certain types of cells on invasion by viruses. There is a substantial literature (though not yet directly relating to the Golgi apparatus) that such modification greatly alters the surface of the cell. Thus it may well be that modification of the genomic control of the Golgi apparatus is a contributing factor in the development of malignancies. MARX (1974) has reviewed the growing focus on cell surfaces as a factor of concern in carcinogenesis. In the normally functioning intact organism these same characteristics by which the cell maintains and alters associations with other cells may function in the relationship of the cell's activities to its environment thus making for the coordinated functioning, through enzymatic and hormonal mechanisms, of the entire organism.

The activity of the Golgi apparatus in the assembly and processing of standard secretory products essential to the integrity of multicellular organisms has now been well documented. Should the hypothesis that the apparatus plays a focal role in the specialization and characterization of the surface membrane prove correct, study of the precise manner in which this organelle functions and how its functional activities can be modulated or controlled could provide a key to a wide spectrum of biological phenomena.

# Bibliography

AKAI, H., 1971: Ultrastructure of fibroin in the silk gland of larval *Bombyx mori*. Exp. Cell Res. **69**, 219—233.

ALBERSHEIM, P., T. M. JONES, and P. D. ENGLISH, 1969: Biochemistry of the cell wall in relation to infective processes. Ann. Rev. Phytopath. **7**, 171—194.

ALTMAN, J., 1971: Coated vesicles and synaptogenesis. A developmental study in the cerebellar cortex of the rat. Brain Res. **30**, 311—322.

AMINOFF, D., ed., 1970: Blood and tissue antigens. New York: Academic Press.

AMOS, W. B., and A. V. GRIMSTONE, 1968: Intercisternal material in the Golgi body of *Trichomonas*. J. Cell Biol. **38**, 466—471.

AMSTERDAM, A., I. OHAD, and M. SCHRAMM, 1969: Dynamic changes in the ultrastructure of the acinar cell of the rat parotid gland during the secretory cycle. J. Cell Biol. **41**, 753—773.

— M. SCHRAMM, I. OHAD, Y. SALOMON, and Z. SELINGER, 1971: Concomitant synthesis of membrane protein and exportable protein of the secretory granule in rat parotid gland. J. Cell Biol. **50**, 187—200.

ANDERSON, E., 1968: Cortical alveoli formation and vitellogenesis during oocyte differentiation in the pipefish, *Syngnathus fuscus*, and the killifish, *Fundulus heteroclitus*. J. Morph. **125**, 23—60.

ANDERSON, W. A., 1968: Cytochemistry of sea urchin gametes. II. Ruthenium red staining of gamete membranes of sea urchins. J. Ultrastruct. Res. **24**, 322—333.

ARCHER, G. T., and J. G. HIRSCH, 1963 a: Isolation of granules from eosinophil leucocytes and study of their enzyme content. J. exp. Med. **118**, 277—286.

— — 1963 b: Motion picture studies on degranulation of horse eosinophils during phagocytosis. J. exp. Med. **118**, 287—294.

BACA, M., and L. ZAMBONI, 1967: The fine structure of human follicular oocytes. J. Ultrastruct. Res. **19**, 354—381.

BACCETTI, B., ed., 1970: Comparative spermatology. New York: Academic Press.

BAINTON, D. F., 1973: Sequential degranulation of the two types of polymorphonuclear leukocyte granules during phagocytosis of microorganisms. J. Cell Biol. **58**, 249—264.

— and M. G. FARQUHAR, 1966: Origin of granules in polymorphonuclear leukocytes. Two types derived from opposite faces of the Golgi complex in developing granulocytes. J. Cell Biol. **28**, 277—302.

— — 1968 a: Differences in enzyme content of azurophil and specific granules of polymorphonuclear leukocytes. I. Histochemical staining of bone marrow smears. J. Cell Biol. **39**, 286—298.

— — 1968 b: Differences in enzyme content of azurophil and specific granules of polymorphonuclear leukocytes. II. Cytochemistry and electron microscopy of bone marrow cells. J. Cell Biol. **39**, 299—317.

— — 1970: Segregation and packaging of granule enzymes in eosinophilic leukocytes. J. Cell Biol. **45**, 54—73.

BAJER, A., and R. D. ALLEN, 1966: Role of phragmoplast filaments in cell-plate formation. J. Cell Sci. **1**, 455—462.

BAJER, A., and J. MOLÈ-BAJER, 1972: Spindle dynamics and chromosome movements. Int. Rev. Cytol. Suppl. 3, 1—255.

BAKER, J. R., 1944: The structure and chemical composition of the Golgi element. Quart. J. micr. Sci. 85, 1—71.

— 1949: Further remarks on the Golgi element. Quart. J. micr. Sci. 90, 293—307.

— 1953: Golgi bodies. A reply to M. D. SRIVASTAVA. Nature (Lond.) 172, 690.

— 1955: What is the "Golgi controversy?" J. roy. micr. Soc. 74, 217—221.

— 1957: The Golgi controversy. Symp. Soc. exp. Biol. 10, 1—10.

— 1963: New developments in the Golgi controversy. J. roy. micr. Soc. 82, 145—157.

BARGMANN, W., 1966: Neurosecretion. Int. Rev. Cytol. 19, 183—201.

BARLAND, P., C. SMITH, and D. HAMERMAN, 1968: Localization of hyaluronic acid in synovial cells by radioautography. J. Cell Biol. 37, 13—26.

BEAMS, H. W., and R. G. KESSEL, 1968: The Golgi apparatus: structure and function. Int. Rev. Cytol. 23, 209—276.

— — 1969: Synthesis and deposition of oocyte envelopes (vitelline membrane, chorion) and the uptake of yolk in the dragonfly (Odonata: Aeschnidae). J. Cell Sci. 4, 241—264.

— and R. L. KING, 1935 a: The effect of ultracentrifuging on the cells of the root tip of the bean (Phaseolus vulgaris). Proc. roy. Soc. (Lond.) B 118, 264—276.

— — 1935 b: Effect of ultra-centrifuging on the cells of the root-tip of the bean. Nature (Lond.) 135, 232.

BEHNKE, O., and H. MOE, 1964: An electron microscope study of mature and differentiating Paneth cells in the rat, especially of their endoplasmic reticulum and lysosomes. J. Cell Biol. 22, 633—652.

BÉLANGER, L. F., 1954 a: Autoradiographic visualization of $S^{35}$ incorporation and turnover by the mucous glands of the gastro-intestinal tract and other soft tissues of rat and hamster. Anat. Rec. 118, 755—771.

— 1954 b: Autoradiographic visualization of the entry and transit of $S^{35}$ in cartilage, bone, and dentine of young rats and the effect of hyaluronidase in vitro. Canad. J. Biochem. Physiol. 32, 161—169.

BENES, F. M., J. A. HIGGINS, and R. J. BARRNETT, 1972: Fine structural localization of acyltransferase activity in rat hepatocytes. J. Histochem. Cytochem. 20, 1031—1040.

BEN-HAYYIM, G., and I. OHAD, 1965: Synthesis of cellulose by Acetobacter xylinum. VIII. On the formation and orientation of bacterial cellulose fibrils in the presence of acidic polysaccharides. J. Cell Biol. 25 (2/2), 191—207.

BENNETT, D., E. A. BOYSE, and L. J. OLD, 1972: Cell surface immunogenetics in the study of morphogenesis. In: Cell interactions (SILVESTRI, L. G., ed.), pp. 247—263. Amsterdam: North-Holland.

BENNETT, G., 1970: Migration of glycoprotein from Golgi apparatus to cell coat in the columnar cells of the duodenal epithelium. J. Cell Biol. 45, 668—673.

— and C. P. LEBLOND, 1970: Formation of cell coat material for the whole surface of columnar cells in the rat small intestine, as visualized by radioautography with L-fucose-³H. J. Cell Biol. 46, 409—416.

— — and A. HADDAD, 1974: Migration of glycoprotein from the Golgi apparatus to the surface of various cell types as shown by radioautography after labeled fucose injection into rats. J. Cell Biol. 60, 258—284.

BENNETT, H. S., 1963: Morphological aspects of extracellular polysaccharides. J. Histochem. Cytochem. 11, 14—23.

BENSLEY, R. R., 1910: On the nature of the canalicular apparatus of animal cells. Biol. Bull. 19, 179—194.

BERG, N. B., and R. W. YOUNG, 1971: Sulfate metabolism in pancreatic acinar cells. J. Cell Biol. 50, 469—483.

BERGEN, F. VON, 1904: Zur Kenntnis gewisser Strukturbilder („Netzapparate", „Saftkanälchen", „Trophospongien") im Protoplasma verschiedener Zellenarten. Arch. mikrosk. Anat. 64, 498—574.

BERGERON, J. J. M., J. H. EHRENREICH, P. SIEKEVITZ, and G. E. PALADE, 1973: Golgi fractions prepared from rat liver homogenates. II. Biochemical characterization. J. Cell Biol. **59**, 73—88.

BERNFIELD, M. R., and N. K. WESSELLS, 1970: Intra- and extracellular control of epithelial morphogenesis. Dev. Biol. Suppl. **4**, 195—249.

— S. D. BANERJEE, and R. H. COHN, 1972: Dependence of salivary epithelial morphology and branching morphogenesis upon acid mucopolysaccharide-protein (proteoglycan) at the epithelial surface. J. Cell Biol. **52**, 674—689.

BIGGERS, J. D., and A. W. SCHUETZ, eds., 1972: Oogenesis. London: Butterworths.

BIONDI, G., 1911: Sulla fine struttura dell'epitelio dei plessi corodei. Archiv Zellf. **6**, 387—396.

BLOCH, D. P., and H. Y. C. HEW, 1960: Schedule of spermatogenesis in the pulmonate snail *Helix aspersa* with special reference to histone transition. J. biophys. biochem. Cytol. **7**, 515—532.

BLOOM, W., and D. W. FAWCETT, 1968: A textbook of histology, 9th edition. Philadelphia: W. B. Saunders Co.

BOSMANN, H. B., and J. J. JACKSON, 1968: Glycoprotein structure: the carbohydrate of bovine corneal collagen. Biochim. biophys. Acta **170**, 6—14.

BOUCHILLOUX, S., O. CHABAUD, M. MICHEL-BÉCHET, M. FERRAND, and A. M. ATHOUËL-HAON, 1970: Differential localization in thyroid microsomal subfractions of a mannosyltransferase, two N-acetylglucosaminyltransferases and a galactosyltransferase. Biochem. biophys. Res. Commun. **40**, 314—320.

BOUCK, G. B., 1965: Fine structure and organelle associations in brown algae. J. Cell Biol. **26**, 523—538.

— and B. M. SWEENEY, 1966: The fine structure and ontogeny of trichocysts in marine dinoflagellates. Protoplasma **61**, 205—223.

BOURNE, G. H., 1955: Some chemical and biochemical aspects of the Golgi apparatus. J. roy. micr. Soc. **74**, 180—187.

BOWEN, R. H., 1924: On the acrosome of the animal sperm. Anat. Rec. **28**, 1—13.

— 1926: The Golgi apparatus—its structure and functional significance. Anat. Rec. **32**, 151—193.

— 1928: Studies on the structure of plant protoplasm. I. The osmiophilic platelets. Z. Zellforsch. mikroskop. Anat. **6**, 689—725.

— 1929: The cytology of glandular secretion. Quart. Rev. Biol. **4**, 299—324; 484—519.

BOWLES, D. J., and D. H. NORTHCOTE, 1972: The sites of synthesis and transport of extracellular polysaccharides in the root tissues of maize. Biochem. J. **130**, 1133—1145.

BRAMBELL, F. W. R., 1925: II. The part played by the Golgi apparatus in secretion, and its subsequent reformation in the cells of the oviducal glands of the fowl. J. roy. micr. Soc. **45**, 17—30.

BRANTON, D., and H. MOOR, 1964: Fine structure in freeze-etched *Allium cepa* L. root tips. J. Ultrastruct. Res. **11**, 401—411.

BREW, K., 1969: Secretion of α-lactalbumin into milk and its relevance to the organisation and control of lactose synthetase. Nature **222**, 671—672.

BRIMACOMBE, J. S., and J. M. WEBBER, 1964: Mucopolysaccharides. Amsterdam: Elsevier.

BROWN, R. M., JR., 1969: Observations on the relationship of the Golgi apparatus to wall formation in the marine chrysophycean alga, *Pleurochrysis scherffelii* Pringsheim. J. Cell Biol. **41**, 109—123.

— W. HERTH, W. W. FRANKE, and D. ROMANOVICZ, 1973: The role of the Golgi apparatus in the biosynthesis and secretion of a cellulosic glycoprotein in *Pleurochrysis:* A model system for the synthesis of structural polysaccharides. In: Biogenesis of plant cell wall polysaccharides (LOEWUS, F., ed.), pp. 207—257. New York: Academic Press.

— W. W. FRANKE, H. KLEINIG, H. FALK, and P. SITTE, 1969: Cellulosic wall component produced by the Golgi apparatus of *Pleurochrysis scherffelii*. Science **166**, 894—896.

— — — — 1970: Scale formation in chrysophycean algae. I. Cellulosic and noncellulosic wall components made by the Golgi apparatus. J. Cell Biol. **45**, 246—271.

BRUNNGRABER, E. G., 1969 a: The possible role of glycoproteins in neural function. Perspect. Biol. Med. **12**, 467—470.

BRUNNGRABER, E. G., 1969 b: Glycoproteins. In: Handbook of neurochemistry 1 (LAJTHA, A., ed.), pp. 223—244. New York: Plenum Press.

— 1972 a: Biochemistry, function, and neuropathology of the glycoproteins in brain tissue. In: Functional and structural proteins of the nervous system (DAVISON, A. N., P. MANDEL, and I. G. MORGAN, eds.), pp. 109—133. New York: Plenum Press.

— 1972 b: Chemistry and metabolism of glycopeptides derived from brain glycoproteins. In: Glycolipids, glycoproteins, and mucopolysaccharides of the nervous system (ZAMBOTTI, V., G. TETTAMANTI, and M. ARRIGONI, eds.), pp. 17—49. New York: Plenum Press.

BURGER, M. M., 1971 a: Surface changes detected by lectins and implications for growth regulation in normal and in transformed cells. Biomembranes 2, 247—270.

— 1971 b: Cell surfaces in neoplastic transformation. Curr. Topics cell. Reg. 3, 135—193.

BUVAT, R., 1957 a: Formations de Golgi dans les cellules radiculares d'*Allium cepa* L. C. R. Acad. Sci. (Paris) 244, 1401—1403.

— 1957 b: Relations entre l'ergastroplasme et l'appareil vacuolaire. C. R. Acad. Sci. (Paris) 245, 350—352.

— 1958: Recherches sur les infrastructures du cytoplasme dans les cellules du méristème apical, des ébauches foliaires et des feuilles développées d'*Elodea canadensis*. Ann. Sci. Nat. Bot., Serie 11e, 19, 121—161.

CAJAL, S. R., 1908: Les conduits de Golgi-Holmgren du protoplasma nerveux et le réseau pericellulaire de la membrane. Trav. Lab. Rech. biol. Madr. 6, 123—135.

— 1914: Algunas variaciones fisiologicas y patologicas del aparato reticular de Golgi. Trab. Lab. Inv. biol. Madr. 12, 127—227.

— 1923: Recuerdos de mi Vida (3rd edition). Madrid: JUAN PUEYO. (Translation: Recollections of My Life. E. HORNE CRAIGIE and JUAN CANO, Memoirs of the American Philosophical Society Vol. VIII, 1937. Philadelphia: The American Philosophical Society.)

CAPALDI, R. A., 1974: A dynamic model of cell membranes. Sci. Amer. 230 (# 3), 27—33.

CARDELL, R. R., JR., S. BANDENHAUSEN, and K. R. PORTER, 1967: Intestinal triglyceride absorption in the rat. An electron microscopical study. J. Cell Biol. 34, 123—156.

CARO, L., and G. E. PALADE, 1964: Protein synthesis, storage, and discharge in the pancreatic exocrine cell. An autoradiographic study. J. Cell Biol. 20, 473—495.

CARTER, H. E., P. JOHNSON, and E. J. WEBER, 1965: Glycolipids. Ann. Rev. Biochem. 34, 109—142.

CECCARELLI, B., W. P. HURLBUT, and A. MAURO, 1973: Turnover of transmitter and synaptic vesicles at the frog neuromuscular junction. J. Cell Biol. 57, 499—524.

CHAMBERS, R., 1940: The relation of extraneous coats to the organization and permeability of cellular membranes. Cold Spring Harbor Symp. Quant. Biol. 8, 144—153.

CHARDAR, R., and C. ROUILLER, 1957: L'ultrastructure de trois algues desmidiées. Étude an microscope électronique. Rev. Cytol. Biol. Veg. 18, 153—178.

CHRÉTIEN, M., 1972: Action de la testostérone sur la structure fine d'un effecteur: la glande sous-maxillaire de la souris male. II. Réaction des tubes sécréteurs a l'injection de testostérone chez le castrat. J. Microscopie 14, 55—74.

CLAUDE, A., 1970: Growth and differentiation of cytoplasmic membranes in the course of lipoprotein granule synthesis in the hepatic cell. I. Elaboration of elements of the Golgi complex. J. Cell Biol. 47, 745—766.

COGGESHALL, R. E., 1972 a: Autoradiographic and chemical localization of 5-hydroxytryptamine in identified neurons in the leech. Anat. Rec. 172, 489—498.

— 1972 b: The structure of the accessory genital mass in *Aplysia californica*. Tissue and Cell 4, 105—127.

COHN, Z. A., and M. E. FEDORKO, 1969: Lysosomal form and function. In: Lysosomes in biology and pathology 1 (DINGLE, J. T., and H. B. FELL, eds.), pp. 43—63. Amsterdam: North-Holland.

— — and J. G. HIRSCH, 1966 a: The *in vitro* differentiation of mononuclear phagocytes. V. The formation of macrophage lysosomes. J. exp. Med. 123, 757—765.

Cohn, Z. A., J. G. Hirsch, and M. E. Fedorko, 1966 b: The *in vitro* differentiation of mono-nuclear phagocytes. IV. The ultrastructure of macrophage differentiation in the peritoneal cavity and in culture. J. exp. Med. **123**, 747—755.

Colwin, L. H., and A. L. Colwin, 1967: Membrane fusion in relation to sperm-egg association. In: Fertilization **1** (Metz, C. B., and A. Monroy, eds.), pp. 295—367. New York: Academic Press.

Cook, G. M. W., 1973: The Golgi apparatus: form and function. In: Lysosomes in biology and pathology **3** (Dingle, J. T., ed.), pp. 237—277. Amsterdam: North-Holland.

Cooper, G. W., and D. J. Prockop, 1968: Intracellular accumulation of protocollagen and extrusion of collagen by embryonic cartilage cells. J. Cell Biol. **38**, 523—537.

Coulomb, P., and J. Coulon, 1971: Fonctions de l'appareil de Golgi dans les méristèmes radiculaires de la courge (*Cucurbita pepo* L. Cucurbitacée). J. Microscopie **10**, 203—214.

Coulombre, A. J., and J. L. Coulombre, 1972: Corneal development. IV. Interruption of collagen excretion into the primary stroma of the cornea with L-azetidine-2-carboxylic acid. Dev. Biol. **28**, 183—190.

Cowdry, E. V., 1923: The significance of the internal reticular apparatus of Golgi in cellular physiology. Science **58**, 1—7.

— 1924: Section 6: cytological constituents—mitochondria, Golgi apparatus and chromidial substance. In: General cytology (Cowdry, E. V., ed.). Chicago: University of Chicago Press.

Cowie, A. T., and J. S. Tindal, 1971: The Physiology of Lactation. Baltimore: Williams and Wilkins.

Creamer, B., 1967: Paneth cell function. Lancet **1**, 314—316.

— and I. J. Pink, 1967: Paneth cell deficiency. Lancet **1**, 304—306.

Cronshaw, J., and K. Esau, 1968: Cell division in leaves of *Nicotiana*. Protoplasma **65**, 1—24.

Cunningham, W. P., 1974: Isolation of the Golgi apparatus. In: Subcellular particles, structures, and organelles (Laskin, A. I., and J. A. Last, eds.), pp. 111—154. New York: Marcel Dekker, Inc.

— D. J. Morré, and H. H. Mollenhauer, 1966: Structure of isolated plant Golgi apparatus revealed by negative staining. J. Cell Biol. **28**, 169—179.


Da Fano, C., 1926: Camillo Golgi 1843–1926. J. Pathol. Bacteriol. **29**, 500—514.

Dalton, A. J., 1961: Golgi apparatus and secretion granules. In: The cell **2** (Brachet, J., and A. E. Mirsky, eds.), pp. 603—619. New York: Academic Press.

— and M. D. Felix, 1954: Cytologic and cytochemical characteristics of the Golgi substance of epithelial cells of the epididymis—*in situ*, in homogenates, and after isolation. Amer. J. Anat. **94**, 171—208.

— — 1957: Electron microscopy of mitochondria and the Golgi complex. Symp. Soc. exp. Biol. **10**, 148—159.

Dan, J. C., 1967: Acrosome reaction and lysins. In: Fertilization **1** (Metz, C. B., and A. Monroy, eds.), pp. 237—293. New York: Academic Press.

— 1970: Morphogenetic aspects of acrosome formation and reaction. Adv. Morphogen. **8**, 1—39.

Daniels, E. W., 1964: Origin of the Golgi system in amoebae. Z. Zellforsch. mikroskop. Anat. **64**, 38—51.

Dauwalder, M., and W. G. Whaley, 1973: Staining of cells of *Zea mays* root apices with the osmium-zinc iodide and osmium impregnation techniques. J. Ultrastruct. Res. **45**, 279—296.

— — 1974: Patterns of incorporation of [³H]-galactose by cells of *Zea mays* root tips. J. Cell Sci. **14**, 11—27.

— — and J. E. Kephart, 1969: Phosphatases and differentiation of the Golgi apparatus. J. Cell Sci. **4**, 455—497.

— — — 1972: Functional aspects of the Golgi apparatus. Sub-Cell. Biochem. **1**, 225—276.

Dawson, P. A., 1973: Observations on the structure of some forms of *Gomphonema parvulum* Kütz. II. The internal organization. J. Phycol. **9**, 165—175.

DE CASTRO, N. M., W. DA SILVA SASSO, and F. A. SAAD, 1959: Preliminary observations of the Paneth cells of the *Tamandua tetradactyla* Lin. Acta Anat. **38**, 345.

DE DUVE, C., 1963: The lysosome. Sci. Amer. **208** (# 5), 64—72.

— 1969: The lysosome in retrospect. In: Lysosomes in biology and pathology 1 (DINGLE, J. T., and H. B. FELL, eds.), pp. 3—40. Amsterdam: North-Holland.

DEINEKA, D., 1912: Der Netzapparat von Golgi in einigen Epithel- und Bindegewebzellen während der Ruhe und während der Teilung derselben. Anat. Anz. **41**, 289—309.

— 1916: as cited by NASSONOV, 1924.

DEL CASTILLO, J., and B. KATZ, 1954: Quantal components of the end-plate potential. J. Physiol. **124**, 560—573.

— — 1956: Biophysical aspects of neuro-muscular transmission. Progr. Biophys. **6**, 121—170.

DE NATTANCOURT, D., M. DEVREUX, A. BOZZINI, M. CRESTI, E. PACINI, and G. SARFATTI, 1973: Ultrastructural aspects of the self-incompatibility mechanism in *Lycopersicum peruvianum* Mill. J. Cell Sci. **12**, 403—419.

DERMER, G. B., 1968: An autoradiographic and biochemical study of oleic acid absorption by intestinal slices including determinations of lipid loss during preparation for electron microscopy. J. Ultrastruct. Res. **22**, 312—325.

DE ROBERTIS, E., 1964: Histophysiology of synapses and neurosecretion. Oxford: Pergamon Press.

DEWEL, W. C., and W. H. CLARK, JR., 1974: A fine structural investigation of surface specializations and the cortical reaction in eggs of the cnidarian *Bunodosoma cavernata*. J. Cell Biol. **60**, 78—91.

DI BENEDETTA, C., and L. A. CIOFFI, 1972: Glycoproteins during the development of the rat brain. In: Glycolipids, glycoproteins, and mucopolysaccharides of the nervous system (ZAMBOTTI, V., G. TETTAMANTI, and M. ARRIGONI, eds.), pp. 115—124. New York: Plenum Press.

DINGLE, J. T., 1969: The extracellular secretion of lysosomal enzymes. In: Lysosomes in biology and pathology 2 (DINGLE, J. T., and H. B. FELL, eds.), pp. 421—436. Amsterdam: North-Holland.

— ed., 1973: Lysosomes in biology and pathology 3. Amsterdam: North-Holland.

— and H. B. FELL, eds., 1969: Lysosomes in biology and pathology 1, 2. Amsterdam: North-Holland.

DISCHE, Z., 1966: The informational potentials of conjugated proteins. In: Protides of the biological fluids (PEETERS, H., ed.), pp. 1—20. Amsterdam: Elsevier.

DOBBINS, W. O., III, 1966: An ultrastructural study of the intestinal mucosa in congenital β-lipoprotein deficiency with particular emphasis upon the intestinal absorptive cell. Gastroenterology **50**, 195—210.

DOTT, H. M., 1969: Lysosomes and lysosomal enzymes in the reproductive tract. In: Lysosomes in biology and pathology 1 (DINGLE, J. T., and H. B. FELL, eds.), pp. 330—360. Amsterdam: North-Holland.

DOUGLAS, S. H., 1935: A note on the work of V. LA VALETTE ST. GEORGE, the discoverer of the Golgi apparatus and mitochondria of modern cytology. J. roy. micr. Soc. **55**, 28—31.

DOUGLAS, W. W., J. NAGASAWA, and R. SCHULZ, 1971: Electron microscopic studies on the mechanism of secretion of posterior pituitary hormones and significance of micro-vesicles ("synaptic vesicles"): evidence of secretion by exocytosis and formation of microvesicles as a byproduct of this process. Mem. Soc. Endocrinol. **19**, 353—378.

DROZ, B., 1965: Accumulation de protéines nouvellement synthétisées dans l'appareil de Golgi du neurone ; étude radioautographique en microscopie électronique. C. R. Acad. Sci. (Paris) **260**, 320—322.

— 1967 a: Synthése et transfert des protéines cellulaires dans les neurones ganglionnaires ; étude radioautographique quantitative en microscopie électronique. J. Microscopie **6**, 201—223.

— 1967 b: L'appareil de Golgi comme site d'incorporation du galactose-$^3$H dans les neurones ganglionnaires spinaux chez le rat. J. Microscopie **6**, 419—424.

— 1969: Protein metabolism in nerve cells. Int. Rev. Cytol. **25**, 363—390.

Droz, B., and H. L. Koenig, 1969: The turnover of proteins in axons and nerve endings. In: Cellular dynamics of the neuron (Barondes, S. H., ed.), pp. 35—50. New York: Academic Press.

Duesberg, J., 1912: Plastosomen, "Apparato reticolare interno" und Chromidialapparat. Ergebn. Anat. EntwGesch. **20**, 567—916.

— 1914: Trophospongien und Golgischer Binnenapparat. Verhandl. Anat. Gesellsch. (XXVIII Vers.) 11—80.

Edelman, G. M., and C. F. Millette, 1971: Molecular probes of spermatozoan structures. Proc. nat. Acad. Sci. (U.S.) **68**, 2436—2440.

Ehrenreich, J. H., J. J. M. Bergeron, P. Siekevitz, and G. E. Palade, 1973: Golgi fractions prepared from rat liver homogenates. I. Isolation procedure and morphological characterization. J. Cell Biol. **59**, 45—72.

Ekholm, R., and L. E. Ericson, 1968: The ultrastructure of the parafollicular cells of the thyroid gland in the rat. J. Ultrastruct. Res. **23**, 378—402.

Elbein, A. D., and W. T. Forsee, 1973: Studies on the biosynthesis of cellulose. In: Biogenesis of plant cell wall polysaccharides (Loewus, F., ed.), pp. 259—295. New York: Academic Press.

Engster, M. S., and S. C. Brown, 1972: Histology and ultrastructure of the tube foot epithelium in the phanerozonian starfish, *Astropecter*. Tissue & Cell **4**, 503—518.

Ericson, L. E., 1972: Formation and storage of 5-hydroxytryptamine in thyroid parafollicular cells. J. Ultrastruct. Res. **41**, 467—483.

Ericsson, J. L. E., 1969: The correlation of structure with function in the liver. In: The biological basis of medicine **5** (Bittar, E. E., and N. Bittar, eds.), pp. 143—209. London: Academic Press.

Eriksson, L., H. Svensson, A. Bergstrand, and G. Dallner, 1972: Physicochemical and enzymic properties of the endoplasmic reticulum in relation to membrane biogenesis. In: Role of membranes in secretory processes (Bolis, L., R. D. Keynes, and W. Wilbrandt, eds.), pp. 3—23. Amsterdam: North-Holland.

Erlandsen, S. L., and D. G. Chase, 1972 a: Paneth cell function: Phagocytosis and intracellular digestion of intestinal microorganisms. I. *Hexamita muris*. J. Ultrastruct. Res. **41**, 296—318.

— — 1972 b: Paneth cell function: Phagocytosis and intracellular digestion of intestinal microorganisms. II. Spiral microorganism. J. Ultrastruct. Res. **41**, 319—333.

Eylar, E. H., 1965: On the biological role of glycoproteins. J. theoret. Biol. **10**, 89—113.

Fañanás, J. R., 1912: Nota preventiva sobre el aparato reticular de Golgi del embrión de pollo. Trab. Lab. Inv. biol. Madr. **10**.

Farquhar, M. G., J. J. M. Bergeron, and G. E. Palade, 1974: Cytochemistry of Golgi fractions prepared from rat liver. J. Cell Biol. **60**, 8—25.

Fawcett, D. W., 1962: Physiologically significant specializations of the cell surface. Circulation **26**, 1105—1125.

— 1966: An atlas of fine structure. Philadelphia: W. B. Saunders Co.

Fedorko, M. E., and J. G. Hirsch, 1966: Cytoplasmic granule formation in myelocytes. An electron microscope radioautographic study on the mechanism of formation of cytoplasmic granules in rabbit heterophilic myelocytes. J. Cell Biol. **29**, 307—316.

Fewer, P., J. Threadgold, and H. Sheldon, 1964: Studies on cartilage: Electron microscopic observations on the autoradiographic localization of $S^{35}$ in cells and matrix. J. Ultrastruct. Res. **11**, 166—172.

Fitton Jackson, S., 1964: Connective tissue cells. In: The Cell **6** (Brachet, J., and A. E. Mirsky, eds.), pp. 387—520. New York: Academic Press.

— 1968: The morphogenesis of collagen. In: Treatise on collagen **2 B** (Gould, B. S., ed.), pp. 1—66. New York: Academic Press.

— 1970: Morphogenetic influences of intercellular macromolecules in cartilage. In: Chemistry and molecular biology of the intercellular matrix **3** (Balazs, E. A., ed.), pp. 1771—1778. New York: Academic Press.

FLEISCHER, B., and S. FLEISCHER, 1971: Comparison of cellular membranes of liver with emphasis on the Golgi complex as a discrete organelle. Biomembranes **2**, 75—94.

FLICKINGER, C. J., 1968 a: The effects of enucleation on the cytoplasmic membranes of *Amoeba proteus*. J. Cell Biol. **37**, 300—315.

— 1968 b: Cytoplasmic alterations in amebae treated with actinomycin D. A comparison with the effects of surgical enucleation. Exp. Cell Res. **53**, 241—251.

— 1969: The development of Golgi complexes and their dependence upon the nucleus in amebae. J. Cell Biol. **43**, 250—262.

— 1971 a: Decreased formation of Golgi bodies in amebae in the presence of RNA and protein synthesis inhibitors. J. Cell Biol. **49**, 221—226.

— 1971 b: Alterations in the Golgi apparatus of amebae in the presence of an inhibitor of protein synthesis. Exp. Cell Res. **68**, 381—387.

— 1972: Influence of inhibitors of energy metabolism on the formation of Golgi bodies in amebae. Exp. Cell Res. **73**, 154—160.

FRANKE, W. W., and U. SCHEER, 1972: Structural details of dictyosomal pores. J. Ultrastruct. Res. **40**, 132—144.

FREY-WYSSLING, A., J. F. LÓPEZ-SÁEZ, and K. MÜHLETHALER, 1964: Formation and development of the cell plate. J. Ultrastruct. Res. **10**, 422—432.

FRIEDMAN, H. I., and R. R. CARDELL, JR., 1972 a: Effects of puromycin on the structure of rat intestinal epithelial cells during fat absorption. J. Cell Biol. **52**, 15—40.

— — 1972 b: Morphological evidence for the release of chylomicra from intestinal absorptive cells. Exp. Cell Res. **75**, 57—62.

FRIEND, D. S., 1965: The fine structure of Brunner's glands in the mouse. J. Cell Biol. **25** (3/1), 563—576.

— 1969: Cytochemical staining of multivesicular body and Golgi vesicles. J. Cell Biol. **41**, 269—279.

— and M. J. MURRAY, 1965: Osmium impregnation of the Golgi apparatus. Amer. J. Anat. **117**, 135—149.

FUCHS, H., 1902: Über das Epithel im Nebenhoden der Maus. Anat. Hefte **19**, 313—347.


GATENBY, J. B., 1919: The cytoplasmic inclusions of the germ-cells Part V. The gametogenesis and early development of *Limmnaea stagnalis* (L.), with special reference to the Golgi apparatus and the mitochondria. Quart. J. micr. Sci. **63**, 445—491.

— 1955: The Golgi apparatus. J. roy micr. Soc. **74**, 134—161.

GERSHON, M. D., and E. A. NUNEZ, 1973: Subcellular storage organelles for 5-hydroxytrypt-amine in parafollicular cells of the thyroid gland. The effect of drugs which deplete the amine. J. Cell Biol. **56**, 676—689.

GINSBURG, A., and E. R. STADTMAN, 1970: Multienzyme systems. Ann. Rev. Biochem. **39**, 429—472.

GODMAN, G. C., and N. LANE, 1964: On the site of sulfation in the chondrocyte. J. Cell Biol. **21**, 353—366.

— and K. R. PORTER, 1960: Chondrogenesis, studied with the electron microscope. J. biophys. biochem. Cytol. **8**, 719—760.

GOLGI, C., 1898: Sur la structure des cellules nerveuses. Arch. ital. Biol. **30**, 60—71. (Originally published in Boll. Soc. med.-chir. di Pavia, 1898.)

— 1908: Di un metodo per la facile e pronta dimostrazione dell'apparato reticolare interno delle cellule nervose. Boll. Soc. med.-chir. di Pavia, 82—87.

— 1909: Sur une fine particularité de structure de l'epithelium de la muqueuse gastrique et intestinale de quelques vertebrés. Arch. ital. Biol. **51**.

— 1903—1929: Opera Omnia. 4 vols. Milan: Hoepli.

GOOD, R. A., and B. W. PAPERMASTER, 1964: Ontogeny and phylogeny of adaptive immunity. Adv. Immunol. **4**, 1—115.

GOODMAN, E. M., and H. P. RUSCH, 1970: Ultrastructural changes during spherule formation in *Physarum polycephalum*. J. Ultrastruct. Res. **30**, 172—183.

GORDON, A. H., 1973: The role of lysosomes in protein catabolism. In: Lysosomes in biology and pathology 3 (DINGLE, J. T., ed.), pp. 89—137. Amsterdam: North-Holland.

GRANT, P. T., T. L. COOMBS, N. W. THOMAS, and J. R. SARGENT, 1971: The conversion of [$^{14}$C]proinsulin to insulin in isolated subcellular fractions of fish islet preparations. In: Subcellular organization and function in endocrine tissues (HELLER, H., and K. LEDERIS, eds.), pp. 481—493. Cambridge: Cambridge University Press.

GRASSÉ, P.-P., 1957: Ultrastructure, polarité et reproduction de l'appareil de Golgi. C. R. Acad. Sci. (Paris) **245**, 1278—1281.

GRAY, E. G., 1970: The question of relationship between Golgi vesicles and synaptic vesicles in *Octopus* neurons. J. Cell Sci. **7**, 189—201.

GRIFFITH, D. L., and W. BONDAREFF, 1973: Localization of thiamine pyrophosphatase in synaptic vesicles. Amer. J. Anat. **136**, 549—556.

GROVE, S. N., C. E. BRACKER, and D. J. MORRÉ, 1968: Cytomembrane differentiation in the endoplasmic reticulum-Golgi apparatus-vesicle complex. Science **161**, 171—173.

GUILLIERMOND, A., 1929: The recent development of our idea of a vacuome of plant cells. Amer. J. Bot. **16**, 1—22.

— and G. MANGENOT, 1922: Sur la signification de l'appareil réticulaire de Golgi. C. R. Acad. Sci. (Paris) **174**, 692—694.

HADDAD, A., M. D. SMITH, A. HERSCOVICS, N. J. NADLER, and C. P. LEBLOND, 1971: Radioautographic study of *in vivo* and *in vitro* incorporation of fucose-$^3$H into thyroglobulin by rat thyroid follicular cells. J. Cell Biol. **49**, 856—882.

HAKOMORI, S., 1971: Glycolipid changes associated with malignant transformation. In: The dynamic structure of cell membranes (WALLACH, D. F. H., and H. FISCHER, eds.), pp. 65—96. New York: Springer.

HALBHUBER, K.-J., A. CHRISTNER, and W. SCHIRRMEISTER, 1972: Autoradiographische und ultrahistochemische Beobachtungen über die Funktion des Golgi-Apparates. Acta Histochem. **42**, 157—161.

HALL, W. T., and E. R. WITKUS, 1964: Some effects on the ultrastructure of the root meristem of *Allium cepa* by 6 aza uracil. Exp. Cell Res. **36**, 494—501.

HARDIN, J. H., and S. S. SPICER, 1971: Ultrastructural localization of dialyzed iron-reactive mucosubstance in rabbit heterophils, basophils, and eosinophils. J. Cell Biol. **48**, 368—386.

HARRISON, G., 1968: Subcellular particles in echinoderm tube feet. II. Class *Holothuroidea*. J. Ultrastruct. Res. **23**, 124—133.

— and D. PHILPOTT, 1966: Subcellular particles in echinoderm tube feet. I. Class *Asteroidea*. J. Ultrastruct. Res. **16**, 537—547.

HAY, E. D., and J. W. DODSON, 1973: Secretion of collagen by corneal epithelium. I. Morphology of the collagenous products produced by isolated epithelia grown on frozen-killed lens. J. Cell Biol. **57**, 190—213.

— and J.-P. REVEL, 1969: Fine structure of the developing avian cornea. In: Monographs in developmental biology (WOLSKY, A., and P. S. CHEN, eds.), pp. 1—144. Basel: S. Karger.

HEATH, E. C., 1971: Complex polysaccharides. Ann. Rev. Biochem. **40**, 29—56.

HEBB, C. O., 1959: Chemical agents of the nervous system. Int. Rev. Neurobiol. **1**, 165—193.

HEIDENHAIN, M., 1907: Plasma und Zelle. Part 1. Jena: Fischer.

HEITZ, E., 1957 a: Die Struktur der Chondriosomen und Plastiden im Wurzelmeristem von *Zea mais* und *Vicia faba*. Z. Naturf. **12 b**, 283—286.

— 1957 b: Die strukturellen Beziehungen zwischen pflanzlichen und tierischen Chondriosomen. Z. Naturf. **12 b**, 576—578.

— 1957 c: Über Plasmastrukturen bei *Antirrhinum majus* und *Zea mais*. Z. Naturf. **12 b**, 579—583.

— 1958: Weitere Belege für das gesetzmäßige Vorkommen plasmatischer Lamellensysteme bei Pflanzen und ihre identische Struktur mit dem Golgi-Apparat bei Tieren. Z. Naturf. **13 b**, 663—665.

HELMINEN, H. J., and J. L. E. ERICSSON, 1968 a: Studies on mammary gland involution. I. On the ultrastructure of the lactating mammary gland. J. Ultrastruct. Res. **25**, 193—213.

HELMINEN, H. J., and J. L. E. ERICSSON, 1968 b: Studies on mammary gland involution. II. Ultrastructural evidence for auto- and heterophagocytosis. J. Ultrastruct. Res. **25**, 214—227.
— — 1968 c: Studies on mammary gland involution. III. Alterations outside auto- and
heterophagocytic pathways for cytoplasmic degradation. J. Ultrastruct. Res. **25**,
228—239.
— — 1970 a: Quantitation of lysosomal enzyme changes during enforced mammary gland
involution. Exp. Cell Res. **60**, 419—426.
— — 1970 b: On the mechanism of lysosomal enzyme secretion. Electron microscopic and
histochemical studies on the epithelial cells of the rat's ventral prostate lobe. J. Ultrastruct. Res. **33**, 528—549.
— — 1971: Effects of enforced milk stasis on mammary gland epithelium, with special
reference to changes in lysosomes and lysosomal enzymes. Exp. Cell Res. **68**, 411—427.
— — and S. ORRENIUS, 1968: Studies on mammary gland involution. IV. Histochemical
and biochemical observations on alterations in lysosomes and lysosomal enzymes.
J. Ultrastruct. Res. **25**, 240—252.
HEPLER, P. K., and W. T. JACKSON, 1968: Microtubules and early stages of cell-plate
formation in the endosperm of *Haemanthus katherinae* Baker. J. Cell Biol. **38**,
437—446.
HERS, H. G., and F. VAN HOOF, 1969: Genetic abnormalities of lysosomes. In: Lysosomes
in biology and pathology 2 (DINGLE, J. T., and H. B. FELL, eds.), pp. 19—40. Amsterdam:
North-Holland.
HEUSER, J. E., and T. S. REESE, 1973: Evidence for recycling of synaptic vesicle membrane
during transmitter release at the frog neuromuscular junction. J. Cell Biol. **57**, 315—344.
HICKS, R. M., 1966: The function of the Golgi complex in transitional epithelium. Synthesis
of the thick cell membrane. J. Cell Biol. **30**, 623—643.
HIGGINS, J. A., and R. J. BARRNETT, 1971: Fine structural localization of acyltransferases.
The monoglyceride and α-glycerophosphate pathways in intestinal absorptive cells.
J. Cell Biol. **50**, 102—120.
— — 1972: Studies on the biogenesis of smooth endoplasmic reticulum membranes in livers
of phenobarbital-treated rats. I. The site of activity of acyltransferases involved in
synthesis of the membrane phospholipid. J. Cell Biol. **55**, 282—298.
HIRSCHLER, J., 1918: Über den Golgischen Apparat embryonaler Zellen. Arch. mikrosk.
Anat. **91**, 140—181.
HODGE, A. J., J. D. MCLEAN, and F. V. MERCER, 1956: A possible mechanism for the
morphogenesis of lamellar systems in plant cells. J. biophys. biochem. Cytol. **2**,
597—607.
HOFER, H. O., 1968: The phenomenon of neurosecretion. In: The structure and function of
nervous tissue 1 (BOURNE, G. H., ed.), pp. 461—517. New York: Academic Press.
HOFFER, A. P., 1971: The ultrastructure and cytochemistry of the shell membrane-secreting
region of the Japanese quail oviduct. Amer. J. Anat. **131**, 253—288.
HOKIN, L. E., 1968: Dynamic aspects of phospholipids during protein secretion. Int. Rev.
Cytol. **23**, 187—208.
HOLLMANN, K. H., 1969: Quantitative electron microscopy of sub-cellular organization in
mammary gland cells before and after parturition. In: Lactogenesis: The initiation of
milk secretion at parturition (REYNOLDS, M., and S. J. FOLLEY, eds.), pp. 27—41.
Philadelphia: University of Pennsylvania Press.
HOLMGREN, E., 1902: Einige Worte über das "Trophospongium" verschiedener Zellarten.
Anat. Anz. **20**, 433—440.
HOLTZER, H., H. WEINTRAUB, and J. BIEHL, 1973: Cell cycle-dependent events during
myogenesis, neurogenesis, and erythrogenesis. In: Biochemistry of cell differentiation
(MONROY, A., and R. TSANEV, eds.), pp. 41—53. New York: Academic Press.
HOLTZMAN, E., 1969: Lysosomes in the physiology and pathology of neurons. In: Lysosomes
in biology and pathology 1 (DINGLE, J. T., and H. B. FELL, eds.), pp. 192—216.
Amsterdam: North-Holland.
— 1971: Cytochemical studies of protein transport in the nervous system. Phil. Trans.
Roy. Soc. Lond. B **261**, 407—421.

HOLTZMAN, E., A. R. FREEMAN, and L. A. KASHNER, 1971: Stimulation dependent alterations in peroxidase uptake by lobster neuromuscular junctions. Science **173**, 733—736.

— A. B. NOVIKOFF, and H. VILLAVERDE, 1967: Lysosomes and GERL in normal and chromatolytic neurons of the rat ganglion nodosum. J. Cell Biol. **33**, 419—435.

HORWITZ, A. L., and A. DORFMAN, 1968: Subcellular sites for synthesis of chondromuco-protein of cartilage. J. Cell Biol. **38**, 358—368.

HOWELL, S. L., and R. B. L. EWART, 1973: Synthesis and secretion of growth hormone in the rat anterior pituitary. II. Properties of the isolated growth hormone storage granules. J. Cell Sci. **12**, 32—35.

— and P. E. LACY, 1971: Biochemical and ultrastructural studies of insulin storage granules and their secretion. In: Subcellular organization and function in endocrine tissues (HELLER, H., and K. LEDERIS, eds.), pp. 469—478. Cambridge: Cambridge University Press.

— and M. WHITFIELD, 1973: Synthesis and secretion of growth hormone in the rat anterior pituitary. I. The intracellular pathway, its time course and energy requirements. J. Cell Sci. **12**, 1—21.

HSU, W. S., 1963: The nuclear envelope in the developing oocytes of the tunicate, *Boltenia villosa*. Z. Zellforsch. mikroskop. Anat. **58**, 660—678.

HYDE, B. B., 1970: Mucilage-producing cells in the seed coat of *Plantago ovata:* developmental fine structure. Amer. J. Bot. **57**, 1197—1206.


ICHIKAWA, A., 1965: Fine structural changes in response to hormonal stimulation of the perfused canine pancreas. J. Cell Biol. **24**, 369—385.

ISRAEL, H. W., M. M. SALPETER, and F. C. STEWARD, 1968: The incorporation of radioactive proline into cultured cells. J. Cell Biol. **39**, 698—715.

ITO, S., and R. J. WINCHESTER, 1963: Fine structure of the gastric mucosa in the bat. J. Cell Biol. **16**, 541—577.


JACKSON, D. S., and J. P. BENTLEY, 1968: Collagen-glycosaminoglycan interactions. In: Treatise on collagen **2 A** (GOULD, B. S., ed.), pp. 189—214. New York: Academic Press.

JAMIESON, J. D., and G. E. PALADE, 1966: Role of the Golgi complex in the intracellular transport of secretory proteins. Proc. nat. Acad. Sci. (U.S.) **55**, 424—431.

— — 1967 a: Intracellular transport of secretory proteins in the pancreatic exocrine cell. I. Role of the peripheral elements of the Golgi complex. J. Cell Biol. **34**, 577—596.

— — 1967 b: Intracellular transport of secretory proteins in the pancreatic exocrine cell. II. Transport to condensing vacuoles and zymogen granules. J. Cell Biol. **34**, 597—615.

— — 1968 a: Intracellular transport of secretory proteins in the pancreatic exocrine cell. III. Dissociation of intracellular transport from protein synthesis. J. Cell Biol. **39**, 580—588.

— — 1968 b: Intracellular transport of secretory proteins in the pancreatic exocrine cell. IV. Metabolic requirements. J. Cell Biol. **39**, 589—603.

— — 1971 a: Condensing vacuole conversion and zymogen granule discharge in pancreatic exocrine cells: metabolic studies. J. Cell Biol. **48**, 503—522.

— — 1971 b: Synthesis, intracellular transport, and discharge of secretory proteins in stimulated pancreatic exocrine cells. J. Cell Biol. **50**, 135—158.

JENNINGS, M. A., and H. W. FLOREY, 1956: Autoradiographic observations on the mucous cells of the stomach and intestine. Quart. J. exp. Physiol. **41**, 131.

JERSILD, R. A., JR., 1966: A radioautographic study of glyceride synthesis *in vivo* during intestinal absorption of fats and labeled glucose. J. Cell Biol. **31**, 413—427.

JONES, A. L., N. B. RUDERMAN, and M. G. HERRERA, 1967: Electron microscopic and biochemical study of lipoprotein synthesis in the isolated perfused rat liver. J. Lipid Res. **8**, 429—466.

JONES, E. A., 1972: Studies on the particulate lactose synthetase of mouse mammary gland and the role of α-lactalbumin in the initiation of lactose synthesis. Biochem. J. **126**, 67—78.

KATZ, B., 1962: The transmission of impulses from nerve to muscle, and the subcellular unit of synaptic action. Proc. roy. Soc. (Lond.) B **155**, 455—477.

KAYE, J. S., 1962: Acrosome formation in the house cricket. J. Cell Biol. **12**, 411—431.

KEENAN, T. W., D. J. MORRÉ, and R. D. CHEETHAM, 1970 a: Lactose synthesis by a Golgi apparatus fraction from rat mammary gland. Nature (Lond.) **228**, 1105—1106.

— — D. E. OLSON, W. N. YUNGHANS, and S. PATTON, 1970 b: Biochemical and morphological comparison of plasma membrane and milk fat globule membrane from bovine mammary gland. J. Cell Biol. **44**, 80—93.

KESSEL, R. G., 1968: Annulate lamellae. J. Ultrastruct. Res. Suppl. **10**, 5—82.

— 1971: Origin of the Golgi apparatus in embryonic cells of the grasshopper. J. Ultrastruct. Res. **34**, 260—275.

— 1973: Structure and function of the nuclear envelope and related cytomembranes. Prog. Surf. Memb. Sci. **6**, 243—329.

— and H. W. BEAMS, 1965: An unusual configuration of the Golgi complex in pigment-producing "test" cells of the ovary of the tunicate, *Styela*. J. Cell Biol. **25** (1/1), 55—67.

KIRKMAN, E., and A. E. SEVERINGHAUS, 1938 a: A review of the Golgi apparatus. Part I. Anat. Rec. **70**, 413—431.

— — 1938 b: A review of the Golgi apparatus. Part II. Anat. Rec. **70**, 557—573.

— — 1938 c: A review of the Golgi apparatus. Part III. Anat. Rec. **71**, 79—103.

KIRSHNER, N., and A. G. KIRSHNER, 1971: Chromogranin A, dopamine β-hydroxylase and secretion from the adrenal medulla. Phil. Trans. Roy. Soc. (Lond.) B **261**, 279—289.

KJOSBAKKEN, J., and J. R. COLVIN, 1973: Biosynthesis of cellulose by a particulate enzyme system from *Acetobacter xylinum*. In: Biogenesis of plant cell wall polysaccharides (LOEWUS, F., ed.), pp. 361—371. New York: Academic Press.

KLEINSCHUSTER, S. J., and A. A. MOSCONA, 1972: Interactions of embryonic and fetal neural retina cells with carbohydrate-binding phytoagglutinins: Cell surface changes with differentiation. Exp. Cell Res. **70**, 397—410.

KNOX, R. B., 1973: Pollen wall proteins: pollen-stigma interactions in ragweed and *Cosmos* (Compositae). J. Cell Sci. **12**, 421—443.

KOEHLER, J. K., and W. D. PERKINS, 1974: Fine structure observations on the distribution of antigenic sites on guinea pig spermatozoa. J. Cell Biol. **60**, 789—795.

KOLATSCHEV, A., 1916: Recherches cytologiques sur les cellules nerveuses des Mollusques. Arch. Russes d'Anat., d'Histol., et d'Embr. **1**.

KOLSTER, R., 1913: Über die durch Golgis Arsenik und Cajals Urannitrat-Silbermethode darstellbaren Zellstrukturen. Verhandl. der anat. Gesellschaft. Versamml. in Greifswald, 10. bis 13. Mai.

KOSHER, R. A., J. W. LASH, and R. R. MINOR, 1973: Environmental enhancement of *in vitro* chondrogenesis. IV. Stimulation of somite chondrogenesis by exogenous chondromucoprotein. Dev. Biol. **35**, 210—220.

KRAEMER, P. M., 1971: Complex carbohydrates of animal cells: biochemistry and physiology of the cell periphery. Biomembranes **1**, 67—190.

KUFF, E. L., and A. J. DALTON, 1959: Biochemical studies of isolated Golgi membranes. In: Subcellular particles (HAYASHI, T., ed.), pp. 114—127. New York: The Ronald Press.

LAMPORT, D. T. A., 1970: Cell wall metabolism. Ann. Rev. Pl. Physiol. **21**, 235—270.

LANDIS, S. C., 1973: Ultrastructural changes in the mitochondria of cerebellar Purkinje cells of nervous mutant mice. J. Cell Biol. **57**, 782—797.

LANE, N., L. CARO, L. R. OTERO-VILARDEBÓ, and G. C. GODMAN, 1964: On the site of sulfation in colonic goblet cells. J. Cell Biol. **21**, 339—352.

LARSON, D. A., 1965: Fine-structural changes in the cytoplasm of germinating pollen. Amer. J. Bot. **52**, 139—154.

LASCELLES, A. K., 1969: Immunoglobulin secretion into ruminant colostrum. In: Lactogenesis: The initiation of milk secretion at parturition (REYNOLDS, M., and S. J. FOLLEY, eds.), pp. 131—136. Philadelphia: University of Pennsylvania Press.

LA VALETTE ST. GEORGE, A. J. H., 1865: Über die Genese der Samenkörper. Part 1. Arch. mikrosk. Anat. **1**, 403—414.

— 1867: Über die Genese der Samenkörper. Part 2. Arch. mikrosk. Anat. **2**, 263—273.

LEBLOND, C. P., 1965: What radioautography has added to protein lore. In: The use of radioautography in investigating protein synthesis (LEBLOND, C. P., and K. B. WARREN, eds.), pp. 321—339. (Symp. Int. Soc. Cell Biol. **4**.) New York: Academic Press.

LEE, Y. C., and D. LANG, 1968: D-galactose di- and trisaccharides from the earthworm cuticle collagen. J. biol. Chem. **243**, 677—680.

LINZELL, J. L., and M. PEAKER, 1971: Mechanism of milk secretion. Physiol. Rev. **51**, 564—597.

LIVETT, B. G., L. B. GEFFEN, and R. A. RUSH, 1971: Immunochemical methods for demonstrating macromolecules in sympathetic neurons. Phil. Trans. Roy. Soc. (Lond.) B **261**, 359—361.

LOEB, J., 1901: Experiments on artificial parthenogenesis in annelids (*Chatopterus*) and the nature of the process of fertilization. Amer. J. Physiol. **4**, 423.

LONGO, F. J., and E. ANDERSON, 1969: Cytological events leading to the formation of the two-cell stage in the rabbit: association of the maternally and paternally derived genomes. J. Ultrastruct. Res. **29**, 86—118.

LUCY, J. A., 1969: Lysosomal membranes. In: Lysosomes in biology and pathology **2** (Dingle, J. T., and H. B. FELL, eds.), pp. 313—341. Amsterdam: North-Holland.

LUDFORD, R. J., 1925: Cell organs during secretion in the epididymis. Proc. roy. Soc. (Lond.) B **98**, 354—372.

LUNDQUIST, F., 1969: Hepatic cell metabolism. In: The biological basis of medicine **5** (BITTAR, E. E., and N. BITTAR, eds.), pp. 211—244. New York: Academic Press.

MANTON, I., 1966 a: Observations on scale production in *Prymnesium parvum*. J. Cell Sci. **1**, 375—380.

— 1966 b: Observations on scale production in *Pyramimonas amylifera* Conrad. J. Cell Sci. **1**, 429—438.

— 1967 a: Further observations on the fine structure of *Chrysochromulina chiton* with special reference to the haptonema, "peculiar" Golgi structure and scale production. J. Cell Sci. **2**, 265—272.

— 1967 b: Further observations on scale formation in *Chrysochromulina chiton*. J. Cell Sci. **2**, 411—418.

— and M. PARKE, 1962: Preliminary observations on scales and their mode of origin in *Chrysochromulina polylepis* sp. nov. J. mar. biol. Ass. U. K. **42**, 565—578.

— D. G. RAYNS, H. ETTL, and M. PARKE, 1965: Further observations on green flagellates with scaly flagella: the genus *Heteromastix* Korshikov. J. mar. biol. Ass. U. K. **45**, 241—255.

MARCORA, F., 1912: Über die Histogenese des Zentralnervensystems mit besonderer Rücksicht auf die innere Struktur der Nervenelemente. Folia neuro-biol. **5**.

MARENGHI, G., 1903: Alcune particolarita di struttura e di innervazione della cue del l'ammocoetes branchialis. Rendiconti del R. Ist. Lomb. di Scienze et Lettere.

MARSHALL, R. D., 1972: Glycoproteins. Ann. Rev. Biochem. **41**, 673—702.

MARX, J. L., 1974: Biochemistry of cancer cells: Focus on the cell surface. Science **183**, 1279—1282.

MASSALSKI, A., and G. F. LEEDALE, 1969: Cytology and ultrastructure of the Xanthophyceae. I. Comparative morphology of the zoospores of *Bumilleria sicula* Borzi and *Tribonema vulgare* Pascher. Br. phycol. J. **4**, 159—180.

MASUR, S. K., E. HOLTZMAN, and R. WALTER, 1972: Hormone-stimulated exocytosis in the toad urinary bladder. Some possible implications for turnover of surface membrane. J. Cell Biol. **52**, 211—219.

— — I. L. SCHWARTZ, and R. WALTER, 1971: Correlation between pinocytosis and hydroosmosis induced by neurohypophyseal hormones and mediated by adenosine 3′, 5′-cyclic monophosphate. J. Cell Biol. **49**, 582—594.

MATHEWS, M. B., 1965: The interaction of collagen and acid mucopolysaccharides. A model for connective tissue. Biochem. J. **96**, 710—716.

MATHEWS, M. B., 1967: Macromolecular evolution of connective tissue. Biol. Rev. **42**, 499—551.
— 1970: The interactions of proteoglycans and collagen—model systems. In: Chemistry and molecular biology of the intercellular matrix **2** (BALAZS, E. A., ed.), pp. 1155—1169. New York: Academic Press.

MATILE, P., 1969: Plant lysosomes. In: Lysosomes in biology and pathology **1** (DINGLE, J. T., and H. B. FELL, eds.), pp. 406—430. Amsterdam: North-Holland.

McRORIE, R. A., and W. L. WILLIAMS, 1974: Biochemistry of mammalian fertilization. Ann. Rev. Biochem. **43**, 777—803.

MELDOLESI, J., and D. COVA, 1971: *In vitro* stimulation of enzyme secretion and the synthesis of microsomal membranes in the pancreas of the guinea pig. J. Cell Biol. **51**, 396—404.
— — 1972 a: Composition of cellular membranes in the pancreas of the guinea pig. IV. Polyacrylamide gel electrophoresis and amino acid composition of membrane proteins. J. Cell Biol. **55**, 1—18.
— — 1972 b: Synthesis and interactions of cytoplasmic membranes in the acinar cells of the guinea pig pancreas. In: Role of membranes in secretory processes (BOLIS, L., R. D. KEYNES, and W. WILBRANDT, eds.), pp. 62—71. Amsterdam: North-Holland.
— J. D. JAMIESON, and G. E. PALADE, 1971 a: Composition of cellular membranes in the pancreas of the guinea pig. I. Isolation of membrane fractions. J. Cell Biol. **49**, 109—129.
— — — 1971 b: Composition of cellular membranes in the pancreas of the guinea pig. II. Lipids. J. Cell Biol. **49**, 130—149.
— — — 1971 c: Composition of cellular membranes in the pancreas of the guinea pig. III. Enzymatic activities. J. Cell Biol. **49**, 150—158.

MERCER, E. H., 1962: The evolution of intracellular phospholipid membrane systems. In: Interpretation of ultrastructure **1** (HARRIS, R. J. C., ed.), pp. 369—384. New York: Academic Press.

METZ, C. B., 1967: Gamete surface components and their role in fertilization. In: Fertilization **1** (METZ, C. B., and A. MONROY, eds.), pp. 163—236. New York: Academic Press.

MEYER, K., 1969: Biochemistry and biology of mucopolysaccharides. Amer. J. Med. **47**, 664—672.

MIGNOT, J.-P., 1965: Étude ultrastructurale des Eugléniens: II. A, Dictyosomes et dictyocinèse chez *Distigma proteus* Ehrbg. B, Mastigonèmes chez *Anisonema costatum* Christen. Protistologica **1** (2), 17—22.
— G. BRUGEROLLE, and G. METENIER, 1972: Compléments a l'étude des mastigonèmes des protistes flagellés. Utilisation de la technique de Thiéry pour la mise en évidence des polysaccharides sur coupes fines. J. Microscopie **14**, 327—342.

MILLS, E. S., and Y. J. TOPPER, 1970: Some ultrastructural effects of insulin, hydrocortisone, and prolactin on mammary gland explants. J. Cell Biol. **44**, 310—328.

MÖLLENDORFF, W., 1913: Über Vitalfärbung des Granula in den Schleimzellen des Säugerdarmes. Abhandl. der anat. Gesellsch. auf der Versammlung in Greifswald, 10. bis 13. Mai.

MOLLENHAUER, H. H., 1965 a: An intercisternal structure in the Golgi apparatus. J. Cell Biol. **24**, 504—511.
— 1965 b: Transition forms of Golgi apparatus secretion vesicles. J. Ultrastruct. Res. **12**, 439—446.
— 1971: Fragmentation of mature dictyosome cisternae. J. Cell Biol. **49**, 212—214.
— and D. J. MORRÉ, 1966: Golgi apparatus and plant secretion. Ann. Rev. Pl. Physiol. **17**, 27—46.
— and W. G. WHALEY, 1963: An observation on the functioning of the Golgi apparatus. J. Cell Biol. **17**, 222—225.
— W. EVANS, and C. KOGUT, 1968: Dictyosome structure in *Euglena gracilis*. J. Cell Biol. **37**, 579—583.
— D. J. MORRÉ, and C. TOTTEN, 1973: Intercisternal substances of the Golgi apparatus— Unstacking of plant dictyosomes using chaotropic agents. Protoplasma **78**, 443—459.

MORRÉ, D. J., and H. H. MOLLENHAUER, 1964: Isolation of the Golgi apparatus from plant cells. J. Cell Biol. **23**, 295—305.

Morré, D. J., H. H. Mollenhauer, and C. E. Bracker, 1971: Origin and continuity of Golgi apparatus. In: Origin and continuity of cell organelles (Reinert, J., and H. Ursprung, eds.), pp. 82—126. Berlin-Heidelberg-New York: Springer.

— R. L. Hamilton, H. H. Mollenhauer, R. W. Mahley, W. P. Cunningham, R. D. Cheetham, and V. S. Lequire, 1970: Isolation of a Golgi apparatus-rich fraction from rat liver. I. Method and morphology. J. Cell Biol. **44**, 484—491.

Moscona, A. A., 1971: Embryonic and neoplastic cell surfaces: availability of receptors for Concanavalin A and wheat germ agglutinin. Science **171**, 905—907.

— 1973: Cell aggregation. In: Cell biology in medicine (Bittar, E. E., ed.), pp. 571—591. New York: J. Wiley.

Muir, H., 1969: The structure and metabolism of mucopolysaccharides (glycosaminoglycans) and the problem of the mucopolysaccharidoses. Amer. J. Med. **47**, 673—690.

Murray, J. A., 1898: Contributions to a knowledge of the Nebenkern in the spermatogenesis of Pulmonata—Helix and Arion. Zoöl. Jahrbücher **11**, 427—440.

Nachman, R., J. G. Hirsch, and M. Baggiolini, 1972: Studies on isolated membranes of azurophil and specific granules from rabbit polymorphonuclear leukocytes. J. Cell Biol. **54**, 133—140.

Nadler, N. J., and C. P. Leblond, 1955: The site and rate of formation of thyroid hormone. Brookhaven Symp. Biol. No. **7**, 40—60.

— — and J. Carneiro, 1960: Site of formation of thyroglobulin in mouse thyroid as shown by radioautography with leucine-H³. Proc. Soc. exp. Biol. Med. **105**, 38—41.

— B. A. Young, C. P. Leblond, and B. Mitmaker, 1964: Elaboration of thyroglobulin in the thyroid follicle. Endocrinol. **74**, 333—354.

Nassonov, D. N., 1923: Das Golgische Binnennetz und seine Beziehungen zu der Sekretion. Untersuchungen über einige Amphibiendrüsen. Arch. mikrosk. Anat. **97**, 136—186.

— 1924: Das Golgische Binnennetz und seine Beziehungen zu der Sekretion (Fortsetzung). Arch. mikrosk. Anat. **100**, 433—472.

Negri, A., 1900: Di una fina particolaritá de struttura delle cellule di alcune ghiandole dei mammiferi. Boll. Soc. med.-chir. di Pavia **13-14**, 69—71.

Neutra, M., and C. P. Leblond, 1966 a: Synthesis of the carbohydrate of mucus in the Golgi complex as shown by electron microscope radioautography of goblet cells from rats injected with glucose-H³. J. Cell Biol. **30**, 119—136.

— — 1966 b: Radioautographic comparison of the uptake of galactose-H³ and glucose-H³ in the Golgi region of various cells secreting glycoproteins or mucopolysaccharides. J. Cell Biol. **30**, 137—150.

— — 1969: The Golgi apparatus. Sci. Amer. **220** (# 2), 100—107.

Nevins, D. J., P. D. English, and P. Albersheim, 1967: The specific nature of plant cell wall polysaccharides. Pl. Physiol. **42**, 900—906.

— — — 1968: Changes in cell wall polysaccharide associated with growth. Pl. Physiol. **43**, 914—922.

Nichols, B. A., D. F. Bainton, and M. G. Farquhar, 1971: Differentiation of monocytes. Origin, nature, and fate of their azurophil granules. J. Cell Biol. **50**, 498—515.

Northcote, D. H., 1968: The organization of the endoplasmic reticulum, the Golgi bodies and microtubules during cell division and subsequent growth. In: Plant cell organelles (Pridham, J. B., ed.), pp. 179—197. New York: Academic Press.

— 1969 a: The synthesis and metabolic control of polysaccharides and lignin during the differentiation of plant cells. Essays Biochem. **5**, 89—137.

— 1969 b: Fine structure of cytoplasm in relation to synthesis and secretion in plant cells. Proc. roy. Soc. (Lond.) B **173**, 21—30.

— 1971 a: Organisation of structure, synthesis and transport within the plant during cell division and growth. Symp. Soc. exp. Biol. **25**, 51—69.

— 1971 b: The Golgi apparatus. Endeavour **30** (# 109), 26—33.

— 1972: Chemistry of the plant cell wall. Ann. Rev. Pl. Physiol. **23**, 113—132.

— and D. R. Lewis, 1968: Freeze-etched surfaces of membranes and organelles in the cells of pea root tips. J. Cell Sci. **3**, 199—206.

Northcote, D. H., and J. D. Pickett-Heaps, 1966: A function of the Golgi apparatus in polysaccharide synthesis and transport in the root-cap cells of wheat. Biochem. J. **98**, 159—167.

Novikoff, A. B., E. Essner, and N. Quintana, 1964: Golgi apparatus and lysosomes. Fed. Proc. **23**, 1010—1022.

Novikoff, P. M., A. B. Novikoff, N. Quintana, and J.-J. Hauw, 1971: Golgi apparatus, GERL, and lysosomes of neurons in rat dorsal root ganglia, studies by thick section and thin section cytochemistry. J. Cell Biol. **50**, 859—886.

Orci, L., K. H. Gabbay, and W. J. Malaisse, 1972: Pancreatic beta-cell web: its possible role in insulin secretion. Science **175**, 1128—1130.

— F. Malaisse-Lagae, M. Ravazolla, and M. Amherdt, 1973 a: Exocytosis-endocytosis coupling in the pancreatic beta cell. Science **181**, 561—562.

— A. E. Lambert, Y. Kanazawa, M. Amherdt, C. Rouiller, and A. E. Renold, 1971: Morphological and biochemical studies of B cells of fetal rat endocrine pancreas in organ culture. Evidence for (pro)insulin biosynthesis. J. Cell Biol. **50**, 565—582.

— A. A. Like, M. Amherdt, B. Blondel, Y. Kanazawa, E. B. Marliss, A. E. Lambert, C. B. Wollheim, and A. E. Renold, 1973 b: Monolayer cell culture of neonatal rat pancreas: an ultrastructural and biochemical study of functioning endocrine cells. J. Ultrastruct. Res. **43**, 270—297.

Oseroff, A. R., P. W. Robbins, and M. M. Burger, 1973: The cell surface membrane: biochemical aspects and biophysical probes. Ann. Rev. Biochem. **42**, 647—682.

Ovtracht, L., 1972: Morphologie de la région Golgienne des cellules de la glande multifide de l'escargot au cours du cycle sécrétoire annuel. J. Microscopie **15**, 353—376.

— and J.-P. Thiéry, 1972: Mise en évidence par cytochimie ultrastructurale de compartiments physiologiquement différents dans un même saccule Golgien. J. Microscopie **15**, 135—170.

Palade, G. E., 1956: Intracisternal granules in the exocrine cells of the pancreas. J. biophys. biochem. Cytol. **2**, 417—422.

— 1959: Functional changes in the structure of cell components. In: Subcellular particles (Hayashi, T., ed.), pp. 64—83. New York: The Ronald Press.

— 1966: Structure and function at the cellular level. J. Amer. med. Assoc. **198**, 815—825.

— and A. Claude, 1949 a: The nature of the Golgi apparatus. I. Parallelism between intracellular myelin figures and Golgi apparatus in somatic cells. J. Morphol. **85**, 35—70.

— — 1949 b: The nature of the Golgi apparatus. II. Identification of the Golgi apparatus with a complex of myelin figures. J. Morphol. **85**, 71—112.

Palay, S. L., 1967: Principles of cellular organization in the nervous system. In: The neurosciences (Quarton, G. C., T. Melnechuk, and F. O. Schmitt, eds.), pp. 24—31. New York: The Rockefeller University Press.

— and G. E. Palade, 1955: The fine structure of neurons. J. biophys. biochem. Cytol. **1**, 69—88.

Parat, M., 1928: Contribution à l'étude morphologique et physiologique du cytoplasme. Archs. Anat. microsc. **24**, 73—357.

— and J. Painlevé, 1924 a: Observation vitale d'une cellule glandulaire et activité. Nature et rôle de l'appareil interne de Golgi et de l'appareil de Holmgren. C. R. Acad. Sci. (Paris) **179**, 612—614.

— — 1924 b: Appareil réticulaire interne de Golgi, trophosponge de Holmgren, et vacuome. C. R. Acad. Sci. (Paris) **179**, 844—846.

Parks, H. F., 1962: Unusual formations of ergastoplasm in parotid acinous cells of mice. J. Cell Biol. **14**, 221—234.

Patton, S., 1969: Milk. Sci. Amer. **221** (# 1), 58—68.

— and E. G. Trams, 1971: The presence of plasma membrane enzymes on the surface of bovine milk fat globules. FEBS Lett. **14**, 230—232.

Paul, D. C., and C. W. Goff, 1973: Comparative effects of caffeine, its analogues and calcium deficiency on cytokinesis. Exp. Cell Res. **78**, 399—413.

PERNER, E. S., 1957: Zum elektronenmikroskopischen Nachweis des „Golgi-Apparates" in Zellen höherer Pflanzen. Naturwiss. **44**, 336.

— 1958: Elektronenmikroskopische Untersuchungen zur Cytomorphologie des sogenannten „Golgisystems" in Wurzelzellen verschiedener Angiospermen. Protoplasma **49**, 407—446.

PERRONCITO, A., 1910: Contribution à l'étude de la biologie cellulaire. Mitochondres, chromidies et appareil réticulaire interne dans les cellules spermatiques. Le phénomène de la dictyokinèse. Arch. ital. Biol. **54**, 307—345. (Originally published in Rend. R. Ist. Lomb. Sci. e Let. **16-17**, 1908—1909.)

PERRY, M. M., and C. H. WADDINGTON, 1966: The ultrastructure of the cement gland in *Xenopus laevis*. J. Cell Sci. **1**, 193—200.

PETERSON, M., and C. P. LEBLOND, 1964 a: Synthesis of complex carbohydrates in the Golgi region, as shown by radioautography after injection of glucose. J. Cell Biol. **21**, 143—148.

— — 1964 b: Uptake by the Golgi region of glucose labelled with tritium in the 1 or 6 position, as an indicator of synthesis of complex carbohydrates. Exp. Cell Res. **34**, 420—423.

PHILPOTT, D. E., A. B. CHAET, and A. L. BURNETT, 1966: A study of the secretory granules of the basal disk of Hydra. J. Ultrastruct. Res. **14**, 74—84.

PICKETT-HEAPS, J. D., 1967: The effects of colchicine on the ultrastructure of dividing plant cells, xylem wall differentiation and distribution of cytoplasmic microtubules. Dev. Biol. **15**, 206—236.

PLATNER, G., 1889: Beiträge zur Kenntnis der Zelle und ihrer Teilung. Arch. mikrosk. Anat. **33**, 180—216.

PLATZER, A. C., and S. GLUECKSOHN-WAELSCH, 1972: Fine structure of mutant (muscular dysgenesis) embryonic mouse muscle. Dev. Biol. **28**, 242—252.

PORTER, D., 1969: Ultrastructure of *Labyrinthula*. Protoplasma **67**, 1—19.

PORTER, K. R., 1957: The submicroscopic morphology of protoplasm. Harvey Lectures 1955-1956, **51**, 175—228.

POUX, N., 1963: Localisation de la phosphatase acide dans les cellules méristématiques de Blé (*Triticum vulgare* Vill.). J. Microscopie **2**, 485—490.

— 1965: Localisations de l'activité phosphatasique acide et des phosphates dans les grains d'aleurone. I. Grain d'aleurone renfermant á la fois globoides et cristalloides. J. Microscopie **4**, 771—782.

— 1970: Localisation d'activités dans le méristème radiculaire de *Cucumis sativus* L. III. Activité phosphatasique acide. J. Microscopie **9**, 407—434.

PRICE, D. L., and K. R. PORTER, 1972: The response of ventral horn neurons to axonal transection. J. Cell Biol. **53**, 24—37.

PROCKOP, D. J., O. PETTENGILL, and H. HOLTZER, 1964: Incorporation of sulfate and the synthesis of collagen by cultures of embryonic chondrocytes. Biochim. biophys. Acta. **83**, 189—196.

RAJARAMAN, R., and O. P. KAMRA, 1969: Ultrastructural changes in *Ulva lactuca* Linnaeus after exposure to ruby or neodymium laser radiation: a preliminary report. J. Ultrastruct. Res. **29**, 430—437.

RAMBOURG, A., 1971: Morphological and histochemical aspects of glycoproteins at the surface of animal cells. Int. Rev. Cytol. **31**, 57—114.

— and C. P. LEBLOND, 1967: Electron microscope observations on the carbohydrate-rich cell coat present at the surface of cells in the rat. J. Cell Biol. **32**, 27—53.

— W. HERNANDEZ, and C. P. LEBLOND, 1969: Detection of complex carbohydrates in the Golgi apparatus of rat cells. J. Cell Biol. **40**, 395—414.

— A. MARRAUD, and M. CHRÉTIEN, 1973: Tri-dimensional structure of the forming face of the Golgi apparatus as seen in the high voltage electron microscope after osmium impregnation of the small nerve cells in the semilunar ganglion of the trigeminal nerve. J. Microscopy **97**, 49—57.

— M. NEUTRA, and C. P. LEBLOND, 1966: Presence of a "cell coat" rich in carbohydrate at the surface of all cells in the rat. Anat. Rec. **154**, 41—71.

RAMÓN Y CAJAL, S., see CAJAL, S. R.

REGER, J. F., 1974: The origin and fine structure of cellular processes in spermatozoa of the tick *Dermacentor andersoni*. J. Ultrastruct. Res. **48**, 420—434.

REVEL, J.-P., 1970: Role of the Golgi apparatus of cartilage cells in the elaboration of matrix glycosaminoglycans. In: Chemistry and molecular biology of the intercellular matrix **3** (BALAZS, E. A., ed.), pp. 1485—1502. New York: Academic Press.

— and E. D. HAY, 1963: An autoradiographic and electron microscopic study of collagen synthesis in differentiating cartilage (chondrocytes). Z. Zellforsch. mikroskop. Anat. **61**, 110—141.

REYNOLDS, M., and S. J. FOLLEY, eds., 1969: Lactogenesis: The initiation of milk secretion at parturition. Philadelphia: University of Pennsylvania Press.

ROBERTS, K., and D. H. NORTHCOTE, 1970: The structure of sycamore callus cells during division in a partially synchronized suspension culture. J. Cell Sci. **6**, 299—321.

— — 1972: Hydroxyproline: Observations on its chemical and autoradiographical localization in plant cell wall protein. Planta (Berl.) **107**, 43—51.

RÖHLICH, P., P. ANDERSON, and B. UVNÄS, 1971: Electron microscope observations on compound 48/80-induced degranulation in rat mast cells. Evidence for sequential exocytosis of storage granules. J. Cell Biol. **51**, 465—483.

ROHR, H., 1965: Die Kollagensynthese in ihrer Beziehung zur submikroskopischen Struktur des Osteoblasten. Elektronenmikroskopische-autoradiographische Untersuchung mit tritiummarkiertem Prolin. Virchows Arch. path. Anat. **338**, 342—354.

— J. SCHMALBECK, and A. FELDHEGE, 1967: Elektronenmikroskopische-autoradiographische Untersuchungen über die Eiweißsynthese in der Brunnerschen Drüse der Maus. Z. Zellforsch. mikroskop. Anat. **80**, 183—204.

— U. SEITTER, and J. SCHMALBECK, 1968: Voraussetzungen und derzeitige Grenzen der quantitativen elektronenmikroskopischen Autoradiographie bei Kinetikstudien an Drüsenzellen. Elektronenmikroskopische-autoradiographische Untersuchungen mit $^3$H-Leucin an der Milchdrüse. Z. Zellforsch. mikroskop. Anat. **85**, 376—397.

ROSEMAN, S., 1970: The synthesis of complex carbohydrates by multiglycosyltransferase systems and their potential function in intercellular adhesion. Chem. Phys. Lipids **5**, 270—297.

ROSS, R., 1968: The fibroblast and wound repair. Biol. Rev. **43**, 51—96.

— 1969: Wound healing. Sci. Amer. **220** (# 6), 40—50.

— and E. P. BENDITT, 1962: Wound healing and collagen formation. III. A quantitative radioautographic study of the utilization of proline-H$^3$ in wounds from normal and scorbutic guinea pigs. J. Cell Biol. **15**, 99—108.

— — 1964: Wound healing and collagen formation. IV. Distortion of ribosomal patterns of fibroblasts in scurvy. J. Cell Biol. **22**, 365—389.

— — 1965: Wound healing and collagen formation. V. Quantitative electron microscope radioautographic observations of proline-H$^3$ utilization by fibroblasts. J. Cell Biol. **27**, 83—106.

ROUGIER, M., 1971: Étude cytochimique de la sécrétion des polysaccharides végétaux à l'aide d'une matériel de choix: les cellules de la coiffe de *Zea mays*. J. Microscopie **10**, 67—82.

ROUILLER, C., and E. FAURÉ-FREMIET, 1958: Structure d'un Flagellé chrysomonadien: *Chromulina psammabia*. Exp. Cell Res. **41**, 47—67.

RUBIN, W., M. D. GERSHON, and L. L. ROSS, 1971: Electron microscope radioautographic identification of serotonin-synthesizing cells in the mouse gastric mucosa. J. Cell Biol. **50**, 399—415.

RUTTER, W. J., R. L. PICTET, and P. W. MORRIS, 1973: Toward molecular mechanisms of developmental processes. Ann. Rev. Biochem. **42**, 601—646.


SAGE, J. A., and R. A. JERSILD, JR., 1971: Comparative distribution of carbohydrates and lipid droplets in the Golgi apparatus of intestinal absorptive cells. J. Cell Biol. **51**, 333—338.

SAGER, R., and G. E. PALADE, 1957: Structure and development of the chloroplast in *Chlamydomonas*. J. biophys. biochem. Cytol. **3**, 463—488.

SALPETER, M. M., 1968: H³-proline incorporation into cartilage: electron microscope auto-radiographic observations. J. Morph. **124**, 387—422.

SANDOZ, D., 1970 a: Évolution des ultrastructures au cours de la formation de l'acrosome du spermatozoïde chez la souris. J. Microscopie **9**, 535—558.

— 1970 b: Étude cytochimique des polysaccharides au cours de la spermatogenèse d'un Amphibien anoure: le Discoglosse *Discoglossus pictus* (Otth.). J. Microscopie **9**, 243—262.

— 1972: Variations ultrastructurales de l'appareil de Golgi au cours des divisions cellulaires dans les spermatocytes de souris. J. Microscopie **15**, 225—246.

SANYAL, S., and A. K. BAL, 1973: Comparative light and electron microscopic study of retinal histogenesis in normal and *rd* mutant mice. Z. Anat. Entwickl.-Gesch. **142**, 219—238.

SCHACHTER, H., and L. RODÉN, 1973: The biosynthesis of animal glycoproteins. In: Metabolic conjugation and metabolic hydrolysis 3 (FISHMAN, W. H., ed.), pp. 1—149. New York: Academic Press.

SCHENKEIN, I., and J. W. UHR, 1970: Immunoglobulin synthesis and secretion. I. Biosynthetic studies of the addition of the carbohydrate moieties. J. Cell Biol. **46**, 42—51.

SCHMALBECK, J., and H. ROHR, 1967: Die Mukopolysaccharid-Synthese in ihrer Beziehung zur Eiweißsynthese in der Brunnerschen Drüse der Maus. (Elektronenmikroskopische-autoradiographische Untersuchung mit ³H-Glukose.) Z. Zellforsch. mikroskop. Anat. **80**, 329—344.

SCHMITT, F. O., 1971: Molecular membranology. In: The dynamic structure of cell membranes (WALLACH, D. F. H., and H. FISCHER, eds.), pp. 5—36. New York: Springer.

SCHNEIDER, F. H., A. D. SMITH, and H. WINKLER, 1967: Secretion from the adrenal medulla: biochemical evidence for exocytosis. Br. J. Pharmacol. Chemother. **31**, 94—104.

SCHNEPF, E., 1969: Sekretion und Exkretion bei Pflanzen. Protoplasmatologia VIII, **8**, 1—181.

SCHUEL, H., W. L. WILSON, K. CHEN, and L. LORAND, 1973: A trypsin-like proteinase localized in cortical granules isolated from unfertilized sea urchin eggs by zonal centrifugation. Role of the enzyme in fertilization. Dev. Biol. **34**, 175—186.

SCHUMAKER, V. N., and G. H. ADAMS, 1969: Circulating lipoproteins. Ann. Rev. Biochem. **38**, 113—136.

SCOTT, F. H., 1905: On the metabolism and action of nerve cells. Brain **28**, 506—526.

SEEGMILLER, R., C. C. FERGUSON, and H. SHELDON, 1972 a: Studies on cartilage. VI. A genetically determined defect in tracheal cartilage. J. Ultrastruct. Res. **38**, 288—301.

— F. C. FRASER, and H. SHELDON, 1971: A new chondrodystrophic mutant in mice. Electron microscopy of normal and abnormal chondrogenesis. J. Cell Biol. **48**, 580—593.

— D. O. OVERMAN, and M. N. RUNNER, 1972 b: Histological and fine structural changes during chondrogenesis in micromelia induced by 6-aminonicotinamide. Dev. Biol. **28**, 555—572.

SELZMAN, H. M., and R. A. LIEBELT, 1962: Paneth cell granule of mouse intestine. J. Cell Biol. **15**, 136—139.

SETTERFIELD, G., and S. T. BAYLEY, 1958: Fine structure of plant protoplasm in relation to growth. Pl. Physiol. **33**: suppl. xlvi.

SHAPIRO, B., 1967: Lipid metabolism. Ann. Rev. Biochem. **36**, 247—270.

SIDMAN, R. L., 1970: Cell proliferation, migration, and interaction in the developing mammalian central nervous systems. In: The neurosciences: Second study program (SCHMITT, F. O., ed.), pp. 100—107. New York: Rockefeller University Press.

— 1972: Cell interactions in developing mammalian central nervous system. In: Cell interactions (SILVESTRI, L. G., ed.), pp. 1—13. Amsterdam: North-Holland.

SIEKEVITZ, P., 1972: Biological membranes: the dynamics of their organization. Ann. Rev. Physiol. **34**, 117—140.

SILVEIRA, M., 1967: Formation of structured secretory granules within the Golgi complex in an acoel Turbellarian. J. Microscopie **6**, 95—100.

SITTE, P., 1958: Die Ultrastruktur von Wurzelmeristemzellen der Erbse (*Pisum sativum*). Protoplasma **49**, 447—522.

SJÖSTRAND, F. S., 1968: Ultrastructure and function of cellular membranes. In: The membranes (DALTON, A. J., and F. HAGUENAU, eds.), pp. 151—210. New York: Academic Press.

— and V. HANZON, 1954 a: Membrane structures of cytoplasm and mitochondria in exocrine cells of mouse pancreas as revealed by high resolution electron microscopy. Exp. Cell Res. **7**, 393—414.

— — 1954 b: Ultrastructure of Golgi apparatus of exocrine cells of mouse pancreas. Exp. Cell Res. **7**, 415—429.

SKAER, R. J., 1973: The secretion and development of nematocysts in a siphonophore. J. Cell Sci. **13**, 371—393.

SLAUTTERBACK, D. B., 1963: Cytoplasmic microtubules. I. Hydra. J. Cell Biol. **18**, 367—388.

SMITH, A. D., 1971: Summing up: some implications of the neuron as a secreting cell. Phil. Trans. Roy. Soc. (Lond.) B **261**, 423—437.

SOUZA SANTOS, H., and W. DA SILVA SASSO, 1973: Ultrastructural and histochemical studies on the mucous granules in the tube feet of the starfish *Echinaster brasiliensis*. Experientia **29**, 473—474.

SPICER, S. S., M. W. STALEY, M. G. WETZEL, and B. K. WETZEL, 1967: Acid mucosubstance and basic protein in mouse Paneth cells. J. Histochem. Cytochem. **15**, 225—242.

SPIRO, R. G., 1970: Glycoproteins. Ann. Rev. Biochem. **39**, 599—638.

STAEHELIN, L. A., and O. KIERMAYER, 1970: Membrane differentiation in the Golgi complex of *Micrasterias denticulata* Bréb. visualized by freeze-etching. J. Cell Sci. **7**, 787—792.

STEIN, O., and Y. STEIN, 1967 a: Lipid synthesis, intracellular transport, storage, and secretion. I. Electron microscopic radioautographic study of liver after injection of tritiated palmitate or glycerol in fasted and ethanol-treated rats. J. Cell Biol. **33**, 319—339.

— — 1967 b: Lipid synthesis, intracellular transport, and secretion. II. Electron microscopic radioautographic study of the mouse lactating mammary gland. J. Cell Biol. **34**, 251—263.

— — 1969: Lecithin synthesis, intracellular transport, and secretion in rat liver. IV. A radioautographic and biochemical study of choline-deficient rats injected with choline-$^3$H. J. Cell Biol. **40**, 461—483.

STEWARD, F. C., and K. MÜHLETHALER, 1953: The structure and development of the cell-wall in the Valoniaceae as revealed by the electron microscope. Ann. Bot. **17**, 295—316.

— H. W. ISRAEL, and M. M. SALPETER, 1974: The labeling of cultured cells of *Acer* with [$^{14}$C]proline and its significance. J. Cell Biol. **60**, 695—701.

STOOLMILLER, A. C., and A. DORFMAN, 1967: Mechanism of hyaluronic acid biosynthesis by Group A Streptococcus. Fed. Proc. **26**, 346.

STUTINSKY, F., ed., 1967: Neurosecretion. Berlin-Heidelberg-New York: Springer.

SUSI, R. F., C. P. LEBLOND, and Y. CLERMONT, 1971: Changes in the Golgi apparatus during spermiogenesis in the rat. Amer. J. Anat. **130**, 251—267.

SUTTON, J. S., and L. WEISS, 1966: Transformation of monocytes in tissue culture into macrophages, epithelioid cells, and multinucleated giant cells. J. Cell Biol. **28**, 303—332.


TAMARIN, A., and P. J. KELLER, 1972: An ultrastructural study of the Byssal thread forming system in *Mytilus*. J. Ultrastruct. Res. **40**, 401—416.

TELLO, E. F., 1913: El reticulo de Golgi algunos tumores y en al granuloma experimental-producido por el Kieselgur. Trab. Lab. Invest. biol. Madr. **11**.

TONER, P. G., K. E. CARR, and G. M. WYBURN, 1971: The digestive system—an ultra-structural atlas and review. London: Butterworth & Co. Ltd.

TRELSTAD, R. L., 1970: The Golgi apparatus in chick corneal epithelium: changes in intracellular position during development. J. Cell Biol. **45**, 34—42.

— 1971: Vacuoles in the embryonic chick corneal epithelium, an epithelium which produces collagen. J. Cell Biol. **48**, 689—694.

— and A. J. COULOMBRE, 1971: Morphogenesis of the collagenous stroma in the chick cornea. J. Cell Biol. **50**, 840—858.

TRIER, J. S., V. LORENZSONN, and K. GROEHLER, 1967: Pattern of secretion of Paneth cells of the small intestine of mice. Gastroentrology **53**, 240—249.

TROUGHTON, W. D., and J. S. TRIER, 1969: Paneth and goblet cell renewal in mouse duodenal crypts. J. Cell Biol. **41**, 251—268.

TURNER, F. R., and W. G. WHALEY, 1965: Intercisternal elements of the Golgi apparatus. Science **147**, 1303—1304.

TYLER, A., 1967: Introduction: problems and procedures of comparative gametology and syngamy. In: Fertilization **1** (METZ, C. B., and A. MONROY, eds.), pp. 1—26. New York: Academic Press.

UHR, J. W., 1970: Intracellular events underlying synthesis and secretion of immunoglobulin. Cell. Immunol. **1**, 228—244.

VIAN, B., and J.-C. ROLAND, 1972: Différenciation des cytomembranes et renouvellement du plasmalemme dans les phénomènes de sécrétions végétales. J. Microscopie **13**, 119—136.

WALKER, G., 1970: The histology, histochemistry and ultrastructure of the cement apparatus of three adult sessile barnacles, *Elminius modestus, Balanus balanoides,* and *Balanus hameri.* Marine Biology **7**, 239—248.

WALNE, P. L., 1967: The effects of colchicine on cellular organization in *Chlamydomonas.* II. Ultrastructure. Amer. J. Bot. **54**, 564—577.

WEIER, T. E., 1932: A comparison of the plastid with the Golgi zone. Biol. Bull. **62**, 126—139.

WEIGL, R., 1912: Vergleichend-zytologische Untersuchungen über den Golgi, Kopsch'schen Apparat und dessen Verhältnis zu anderen Strukturen in den somatischen Zellen und Geschlechtszellen verschiedener Tiere. Bull. de l'Acad. des Sciences de Cracov Mai 1912 Cracovie (Crakow).

WEINSTOCK, A., 1972: Matrix development in mineralizing tissues as shown by radioautography: formation of enamel and dentin. In: Developmental aspects of oral biology (SLAVKIN, H. C., and L. A. BAVETTA, eds.), pp. 201—242. New York: Academic Press.

— and C. P. LEBLOND, 1971: Elaboration of the matrix glycoprotein of enamel by the secretory ameloblasts of the rat incisor as revealed by radioautography after galactose-[3]H injection. J. Cell Biol. **51**, 26—51.

— and R. W. YOUNG, 1972: Synthesis and secretion of S[35]-labeled glycosaminoglycans by osteoblasts and ameloblasts. Anat. Rec. **172**, 424 (Abstr.).

WEINSTOCK, M., and C. P. LEBLOND, 1974: Synthesis, migration, and release of precursor collagen by odontoblasts as visualized by radioautography after [³H]proline administration. J. Cell Biol. **60**, 92—127.

WELLINGS, S. R., 1969: Ultrastructural basis of lactogenesis. In: Lactogenesis: The initiation of milk secretion at parturition (REYNOLDS, M., and S. J. FOLLEY, eds.), pp. 5—25. Philadelphia: University of Pennsylvania Press.

— and J. R. PHILP, 1964: The function of the Golgi apparatus in lactating cells of the BALB/cCrgl mouse. Z. Zellforsch. mikroskop. Anat. **61**, 871—882.

WETZEL, B. K., S. S. SPICER, and S. H. WOLLMAN, 1965: Changes in fine structure and acid phosphatase localization in rat thyroid cells following thyrotropin administration. J. Cell Biol. **25** (3/1), 593—618.

WHALEY, W. G., 1966: Proposals concerning replication of the Golgi apparatus. In: Organisation der Zelle. III. Probleme der biologischen Reduplikation (SITTE, P., ed.), pp. 340—371. Berlin-Heidelberg-New York: Springer.

— 1968: The Golgi apparatus. In: The biological basis of medicine **1** (BITTAR, E. E., and N. BITTAR, eds.), pp. 179—208. New York: Academic Press.

— and H. H. MOLLENHAUER, 1963: The Golgi apparatus and cell plate formation—a postulate. J. Cell Biol. **17**, 216—221.

— M. DAUWALDER, and J. E. KEPHART, 1966: The Golgi apparatus and an early stage in cell plate formation. J. Ultrastruct. Res. **15**, 169—180.

— — 1971: Assembly, continuity, and exchanges in certain cytoplasmic membrane systems. In: Origin and continuity of cell organelles (REINERT, J., and H. URSPRUNG, eds.), pp. 1—45. Berlin-Heidelberg-New York: Springer.

— — — 1972: Golgi apparatus: influence on cell surfaces. Science **175**, 596—599.

WHALEY, W. G., J. E. KEPHART, and H. H. MOLLENHAUER, 1959: Developmental changes in the Golgi apparatus of maize root cells. Amer. J. Bot. **46**, 743—751.

— — — 1964: The dynamics of cytoplasmic membranes during development. In: Cellular membranes in development (LOCKE, M., ed.), pp. 135—173. New York: Academic Press.

WHETSELL, W. O., JR., and R. P. BUNGE, 1969: Reversible alterations in the Golgi complex of cultured neurons treated with an inhibitor of active Na and K transport. J. Cell Biol. **42**, 490—500.

WHUR, P., A. HERSCOVICS, and C. P. LEBLOND, 1969: Radioautographic visualization of the incorporation of galactose-$^3$H and mannose-$^3$H by rat thyroids *in vitro* in relation to the stages of thyroglobulin synthesis. J. Cell Biol. **43**, 289—311.

WILD, A. E., 1973: Transport of immunoglobulins and other proteins from mother to young. In: Lysosomes in biology and pathology 3 (DINGLE, J. T., ed.), pp. 169—215. Amsterdam: North-Holland.

WILSON, E. B., 1925: The cell in development and heredity, 3rd edition. New York: The Macmillan Company.

WILSON, H. V., 1907: On some phenomena of coalescence and regeneration in sponges. J. exp. Zool. **5**, 245—258.

WINKLER, H., 1971: The membrane of the chromaffin granule. Phil. Trans. Roy. Soc. (Lond.) B **261**, 293—303.

— H. HÖRTNAGL, H. HÖRTNAGL, and A. D. SMITH, 1970: Membranes of the adrenal medulla. Characterization of insoluble proteins of chromaffin granules by gel electrophoresis. Biochem. J. **118**, 303—310.

WOESSNER, J. F., JR., 1969: The physiology of the uterus and mammary gland. In: Lysosomes in biology and pathology 1 (DINGLE, J. T., and H. B. FELL, eds.), pp. 299—329. Amsterdam: North-Holland.

WONDRAK, G., 1967: Die exoepithelialen Schleimdrüsenzellen von *Arion empiricorum* (Fér.). Z. Zellforsch. mikroskop. Anat. **76**, 287—294.

WOODING, F. B. P., 1973: Formation of the milk fat globule membrane without participation of the plasmalemma. J. Cell Sci. **13**, 221—235.

WRATHALL, J. R., C. OLIVER, S. SILAGI, and E. ESSNER, 1973: Suppression of pigmentation in mouse melanoma cells by 5-bromodeoxyuridine. Effects on tyrosinase activity and melanosome formation. J. Cell Biol. **57**, 406—423.

WRIGHT, D. G., and S. E. MALAWISTA, 1972: The mobilization and extracellular release of granular enzymes from human leukocytes during phagocytosis. J. Cell Biol. **53**, 788—797.


YASUZUMI, G., 1974: Electron microscope studies on spermiogenesis in various animal species. Int. Rev. Cytol. **37**, 53—119.

YOO, B. Y., 1970: Ultrastructural changes in cells of pea embryo radicles during germination. J. Cell Biol. **45**, 158—171.

YOUNG, R. W., 1973: The role of the Golgi complex in sulfate metabolism. J. Cell Biol. **57**, 175—189.


ZAGURY, D., J. W. UHR, J. D. JAMIESON, and G. E. PALADE, 1970: Immunoglobulin synthesis and secretion. II. Radioautographic studies of sites of addition of carbohydrate moieties and intracellular transport. J. Cell Biol. **46**, 52—63.

ZAMBONI, L., 1971: Fine morphology of mammalian fertilization. New York: Harper & Row.

ZYLSTRA, U., 1972: Histochemistry and ultrastructure of the epidermis and the subepidermal gland cells of the freshwater snails *Lymnaea stagnalis* and *Biomphalaria pfeifferi*. Z. Zellforsch. mikroskop. Anat. **103**, 93—134.

# Author Index

# Subject Index

(Numbers in boldface type refer to pages on which illustrations appear)